Interior Interruptions

Interior Interruptions examines the role of the 'palimpsest' and its relationship to narrative, sustainability, renovation and adaptive reuse. By exploring storytelling, palimpsestic characteristics and techniques, the book argues that these devices play a central role in the consideration of the designed interior.

Narrative has a burgeoning relationship with the palimpsest and this approach embraces an aesthetic of incompleteness and imperfection as a site rich response. It recognises the ongoing 'biography' or heritage of a building as a form of transient architectural narrative that encourages reuse through the continual process of writing, rewriting, overwriting and unwriting. This process has sustainable, societal, archaeological and textual connotations that can be interpreted as a process of 'layering' whereby the architectural shell is viewed as a container; a rich repository that is 'overlain' by surface changes, documents architectural and spatial modifications, and is populated by interior fixtures and fittings that all unite to create an ever-changing interior story.

Exploring case studies from the UK, Netherlands, Palestine, Belgium, Singapore, Spain, Portugal, France, Germany, Brazil, Japan, USA and China and beautifully illustrated in full colour, this book proposes that the act of interior renovation can be viewed as a perpetual form of revisionary storytelling re-imagined as a series of temporal interior 'interruptions'. It is essential reading for students and professionals interested in the built environment, including, but not limited to, interior design, interior decoration, interior architecture and architecture.

Jean Whitehead lives in the South West of England, has published extensively and has over 30 years' experience in Higher Education, both within the UK and abroad. Jean is the recipient of the 'Visionary Tutor' award for Interior Design from the Society of British & International Design (SBID) and is the author of *Creating Interior Atmosphere: Mise-en-scène and Interior Design*.

JEAN WHITEHEAD

Interior Interruptions
Rehabilitating the Old to Design the New

LONDON AND NEW YORK

Designed cover image: HUB Flat, Madrid, Spain - Churtichaga + Quadra-Salcedo,
Photo by Elena Almagro supplied by architects

First published 2025
by Routledge
4 Park Square, Milton Park, Abingdon, Oxon OX14 4RN

and by Routledge
605 Third Avenue, New York, NY 10158

Routledge is an imprint of the Taylor & Francis Group, an informa business

© 2025 Jean Whitehead

The right of Jean Whitehead to be identified as author of this work has been asserted in accordance with sections 77 and 78 of the Copyright, Designs and Patents Act 1988.

All rights reserved. No part of this book may be reprinted or reproduced or utilised in any form or by any electronic, mechanical, or other means, now known or hereafter invented, including photocopying and recording, or in any information storage or retrieval system, without permission in writing from the publishers.

Trademark notice: Product or corporate names may be trademarks or registered trademarks, and are used only for identification and explanation without intent to infringe.

British Library Cataloguing-in-Publication Data
A catalogue record for this book is available from the British Library

ISBN: 978-1-032-35301-2 (hbk)
ISBN: 978-1-032-35300-5 (pbk)
ISBN: 978-1-003-32626-7 (ebk)

DOI: 10.4324/9781003326267

Typeset in Joanna
by Apex CoVantage, LLC

For my Mum and Dad, for always believing in me and for Alan, for his unwavering support.

Contents

Acknowledgements ix
Preface x

Interior Interruptions: An Introduction **One** 1
Interior Interruptions – A Synopsis 1
The Palimpsest and Sustainability 7
The Palimpsest and Narrative 11

What is a Palimpsest? **Two** 28
What is a Palimpsest? 29
Palimpsests: Types, Differences and Commonalities 31

Palimpsest Techniques, Overwriting **Three** 51
An Introduction 52
The 'type' of Interruption – Overwriting 54

Palimpsest Techniques, Unwriting **Four** 87
An Introduction 88
The 'type' of Interruption – Unwriting 88

Palimpsest Techniques, Redrafting or from Interruptions to Disruptions **Five** 123
An Introduction 124
The 'type' of Interruption – Redrafting, from Interruptions to Disruptions 125

Interior Interruptions: A Conclusion **Six** 159
Overview 159
Interior Interruptions – A Summation 161

Index 167

Acknowledgements

I would like to thank unreservedly the many individuals, too numerous to mention, who answered my many emails, kindly provided project images and generally helped to keep this book on course and on schedule. A special mention must go to William Mann, Bethany Rolston, Pedro Pegenaute, Dawn Hepburn, Lucy Qian, Rodrigo Rossi, Louise de Brabander, Cate O'Toole, Amanda de Beaufort, Joan Fontbernat, Balthazar Pothier, Hans Fonk and Joy Seah. For everyone who either made time in their busy schedules to proffer project insights online or kindly provided written responses – specifically Mireia Luzárraga, Kate Darby, David Connor, Ricardo Flores, Eva Prats, Arne Vande Capelle, Jo Taillieu, Neri&Hu, Karine Chartier, Diego Cisi, Josemaría de Churtichaga, Marianna Schmidt, Colin Seah and David Dworkind – once again, thank you. This book is much better because of your generosity, patience and involvement. It would be remiss of me not to thank my many colleagues at Falmouth University, especially Fay Freeman for her patience and forbearance whilst I was writing this book. Thanks must also go to Professor Kevin Singh for his early support alongside the staff at Routledge for their astute editorial guidance and unwavering loyalty, especially Fran Ford and Hannah Studd, Nick Craggs for all his help graphically in bringing the book to life, whilst Patricia Teasdale is a copyeditor of par excellence – I could not have done it without you. Finally, and most importantly, thanks to Alan for once again sharing with me the highs and lows of writing a book. This is the result . . .

Preface

This book is both an accumulation and synthesis of a lifetime of teaching, representing a continual engagement with interiority in all of its fascinating forms. This luxurious position has enabled a degree of introspection relative to the burgeoning theoretical debate that currently surrounds interior architecture and adaptive reuse. A fascination with old buildings from a young age began by accompanying my father to houses he was renovating, an allure that has only grown and matured over the years. These stalwarts of many of our streets, towns or cityscapes have immense appeal for me personally, as mature expressions of an architectural sensibility redolent of an interior atmosphere that has only deepened over time. This temporal and historical legacy ensures an experiential engagement that acknowledges the many lives lived, alterations made and even revenants uncovered that can potentially enrich any future occupation. The desire to seek responsive solutions to this architectural inheritance seemed a logical beginning.

In an era of climate crisis and the need to limit material and energy consumption, important questions need addressing with reference to how existing building stock as a resource should be reused. By exploring the 'palimpsest' as an evolutionary device that encourages change by simultaneously looking to the past and the present, these 'Janus-like' proclivities presented an opportunity to develop a theoretical proposition that could potentially contribute to the debate surrounding reuse in a meaningful manner. Historically, the notion of the palimpsest is inherently not a new premise (you just have to acknowledge the writings of Machado and Robert, or more recently Stone, to realise this), however this study does aspire to a holistic overview that borders on the obsessive. By collecting and collating writings and precedents from a wide range of sources, a cross-disciplinary investigation ensued that reinvigorated this palimpsestic inquisitorial stance. It was an approach that was equally informed by a return to the original manuscript, or 'codex' to use the correct terminology, and the notion of under and overwriting, even unwriting, alongside the recognition of a building's biography as a vehicle for gaining further insights into the palimpsest. This book is the result of this enquiry . . .

Interior Interruptions: An Introduction

One

Figure 1.1 Interrupted interiors embrace both the past and the present in the creation of an ongoing, continually recycled life that prioritises reuse creatively and experientially   RYÙ Peel, Montreal, Canada by Ménard Dworkind Architecture & Design (MRDK).
Copyright: Ménard Dworkind Architecture & Design.

INTERIOR INTERRUPTIONS – A SYNOPSIS

If you are interested in the design of interiors, have ever wondered how you can harness creative inspiration, are intrigued by the notion of a site-specific response and have a desire to design more sustainably, then exploring the concept of a palimpsest is both timely and relevant. By introducing and examining the role of the '**palimpsest**' theoretically, this book aims to establish its 'central' relationship to storytelling, sustainability, building reuse and the **interrupted interior**. In fact, this palimpsestic appraisal will serve to illustrate the value placed on the architectural

DOI: 10.4324/9781003326267-1

presence, history and cultural resonance of buildings, an approach that actively challenges building obsolescence. This is a propitious investigation as the Royal Institute of British Architects (RIBA) '*believes that all new buildings need to achieve net zero whole life carbon by 2030*'[1] (Clark, 2019: 19), but, if '*new buildings only account for 1% of the total UK building stock annually*'[2] (Clark, 2019: 24) then how existing buildings are reused becomes a priority. The British Institute of Interior Design's (BIID) Sustainable Specifying Guide starkly states that '*the construction industry accounts for around 40% of the UK's entire carbon footprint*'[3] (2021: 01); a daunting statistic that is unfortunately echoed globally. Therefore, as an ethical architect or designer who is rightly concerned for the future of the planet, can you make a difference? Considering how and what you 'specify' (specification relates to the choice of materials, lighting and furniture, fixtures and equipment, or ff&e, within a scheme) and how you 'respond' to an existing building that is ripe for renovation has historically been a matter of personal conjecture. However, if there was a theoretical system in place that contributed to the debate surrounding adaptive reuse, that acted as a sounding board, that encouraged you to think more sustainably, to assess what you have, what you can retain or reuse, then surely following that would be an appropriate course of action.

This ambition links cyclically to the notion of the palimpsest, so what is a palimpsest and, perhaps more importantly, why is it relevant to the concept of interior interruptions? Put simply, a palimpsest within Antiquity relates to the 'reuse' of an ancient manuscript or tablet, a definition that will be analysed in depth in later chapters. This reuse is often associated with an imperfect cleaning or scraping of the original surface that inadvertently preserves traces of the original text, allowing both the original and any new or later writings to appear simultaneously. It is these traces and their layered coexistence that makes the palimpsest such an enticing prospect, as this is a device that is clearly preoccupied with history. This layering allows texts from different moments in time to occupy the same temporal zone, creating a visually rich form of storytelling as different eras concurrently coincide. By applying this idea to the study of architecture, occupational coexistence resonates with contemporary concerns relating to reduce, reuse and recycle. A palimpsestic interior or building is repositioned from the traditional end-of-life, or limited shelf-life, to one of 'continued' life. An approach that advocates for longevity by embracing a continuous circularity of reuse.

Given this explanation, the palimpsest's parallels with the built environment and the 'life' of a building seem obvious, as this is clearly in favour of and supports a sustainable agenda. In essence, this is about recognising the ongoing 'timeline' of a building as a form of architectural narrative informed by a continual palimpsestic methodology of writing and rewriting. This process has sustainable connotations that can be interpreted as a process of 'layering' whereby the architectural shell is viewed as a repository that is continually 'overlain' by surface changes, modified by architectural adjustments and adaptations and populated by interior fixtures and fittings. This leads to the proposition, an acknowledgement even, that this metamorphic process informs a perpetual form of revisionary storytelling re-imagined as a series of temporal **'interior interruptions'**. By scrutinising these interruptions through the lens of the palimpsest this theoretical stance becomes central to a deeper understanding of the interior and the practice of adaptive reuse. An approach that acknowledges what has gone before; it is attuned to sustainability and is about recognising the debt of the old in sustainably designing the new.

This consideration naturally leads to an obvious starting point, but it is worth stating nonetheless: that interiors rarely exist in isolation. They are generally located within a specific building, contained by the building envelope, informed by the prevailing atmosphere and constrained by their immediate locality or context. These interior 'locators' are tied to a specific place often referred to as its 'genius loci'[4] (Norberg-Schulz, 1984), and this book will illustrate how this symbiotic relationship can be utilised to inform future built incarnations. Central to these many incarnations is the uncovering and celebrating of expressive site narratives as rich stories that inform and enrich future design decisions. Within this context 'narrative' is simply: '*the art, technique, or process of narrating, or of telling a story*'[5] (dictionary.com, n.d.), and any interior can benefit from this place-centric approach as long as it is sensitive to the heritage of past lives and stylistic loves, of the many manifestations and continuous alterations that the built environment can embody.

The notion that an existing building (often referred to as the host) can have a 'life cycle' is a proposition that requires further debate. Life holistically refers to living organisms that are capable of reproduction or metabolism, but a building is clearly inanimate. However, if a building's life cycle is conceived as a 'timeline', a journey that encompasses the initial build through to its current manifestation, then this is a rich vein of enquiry. Whilst some interiors are built, exist for a short period and are then demolished (think pop ups, exhibitions or event spaces), many coexist with and inhabit existing building stock. These could be host buildings that perhaps have a long history culturally within a specific community, are architecturally rich remnants of past epochs or are of an industrial heritage that just needs to adapt to a new function or era. This adaptation ensures their continued relevance, occupation and usage (employing a process often referred to as adaptive reuse). This type of building usually (but not always) has a robust character, is slightly careworn, can be in a state of ruination or abandonment and often contains evidence of its many past incarnations (or series of 'interruptions') along its timeline. Through adaptive reuse an interior can enter into a marriage of convenience with its host, a marriage whereby both parties, via a positive alliance, create a future together. A future that recognises and values a building's 'assets' culturally, economically, sustainably or historically.

Reuse then potentially links to this history, a dynamic history informed by an endless flow of inhabitants, driven by ever changing economic, cultural and societal needs and constraints. In many instances this occupation, marked by the passage of time, becomes woven into the very fabric of the building, litters the interior of the building and can be salvaged or repurposed to astonishing effect. If you can accept that a building has a life cycle, a full life that leaves inevitable traces, then these 'traces' link to place-bound, 'site specific' narratives. This recognition usefully aligns to the central premise of this book; that of narratives burgeoning relationship with the 'palimpsest' as a process that relates to *something bearing the visible traces of an earlier form*'[6] (Stevenson and Waite, 2011: 1031). This revelatory contextually-driven story encompasses a series of temporal 'interruptions', alterations and modifications as traces of the past, expressive traits that are retained in relation to a building's history. However, this is emphatically not about a dogmatic reverence for the past; instead this book sets out to examine how a palimpsestic approach by embracing the 'ongoing biography' of a building actively accepts and exploits change. Instead, creative revisions or 'interruptions' incorporate enticing temporal slippages where the past and the present align and this serves to both explain and justify the title of this book: *Interior Interruptions: Rehabilitating the Old to Create the New*.

Chapter Summaries

This synopsis serves to highlight the book's focus as primarily concerned with an exploration of interior interruptions via the palimpsest as a theoretical framework for creative response, a stance that naturally embraces narrative and a sustainable agenda. Concise chapter summaries will now articulate the book's content explaining its relevance to such specialisms as interior design, interior architecture, art and architecture. Early chapters draw upon broad themes relating to storytelling, narrative and a site's history or 'biography' before clarifying the nature of a palimpsest and its reach across multiple creative disciplines. This comparative analysis develops into an introverted appraisal that aims to establish the typical characteristics of an interior palimpsest via its key traits, concerns and ingredients. Subsequent chapters decipher these ingredients in order to give a detailed overview of the central theoretical argument. This book is intentionally rich in case studies, selecting globally-significant precedents in order to illustrate the main proposition; however, it will deliberately not include all aspects of sustainability (a vast and ever-expanding area of knowledge). Instead, it will capitalise on those sustainable elements that correlate to the notion of a palimpsest and reinforce the central argument in relation to interior interruptions.

Chapter One aims to contextualise temporal interior 'interruptions' in a chapter that takes an expansive view in introducing the main thematic concerns of the book. By providing a holistic overview, both narrative and sustainability will be highlighted at the inception for their centrality to this form of palimpsestic re-imagining. The notion of a site '**biography**' and site '**narratives**' (both fictional and factual, or design fiction and design non-fiction) are introduced, whilst a re-examination of how a palimpsestic approach is both relevant and timely given its ability to embed **sustainable thinking** is undertaken. The main tenets of environmental practice relating to reduce, reuse, recycle, and even make do and mend, remain central to this approach.

This introductory focus evolves as Chapter Two investigates the typical **commonalities** and **characteristics** of a palimpsest and examines its **theoretical role** in relation to the interrupted interior. The palimpsest's persuasive interdisciplinary reach across multiple disciplines (art, literature, cinema, architecture and design) will be scrutinised, culminating in the identification of common recurring traits. This discussion sets the scene for an introduction to the palimpsest within the context of the built environment, concluding with the establishment of the main concerns and preoccupations of an **interior palimpsest**.

Chapters Three, Four and Five examine in detail important aspects of the interior palimpsest and utilise key precedents to illustrate and deepen the reader's understanding of interior interruptions. A sense of how adaption draws upon the twin auspices of both **alteration** and **retention** will be highlighted, as theoretically both are integral to understanding the interior palimpsest. These are interruptions that will be interrogated for their ability to write, rewrite and unwrite site meaning. Chapters Three to Five inclusive examine common **palimpsestic techniques**, utilising textual and editorial analogies developed from an understanding of the palimpsestic process. Chapter Three explores and illustrates the 'additive' technique of **overwriting**, whilst Chapter Four examines the 'subtractive' technique of **unwriting**. Chapter Five investigates the palimpsest's ability to '**redraft**' meaning by implementing evolutionary change in its deployment of such techniques as **interruptions** and **disruptions**. Holistically these five chapters aim to present the key findings in relation to the establishment of an interior palimpsest culminating in Chapter Six's conclusion.

Worth noting for the sake of clarity is the extensive vocabulary in relation to building reuse and it will be helpful to examine these terms in relation to interior interruptions and the palimpsest. The following section aims to identify commonalities of definition but is conscious that there are always exceptions to any given rule given the existence of cultural and industry-wide variations globally. With this caveat in mind, 'adaption' and 'adaptive reuse' remain useful terminologies as they typically highlight an approach that prioritises the reuse of existing buildings by updating them for a new function as an ethical alternative to new build. 'Retrofitting' as a term commonly refers to the 'upgrading' of a building in relation to energy efficiency or essential services, but it can also encompass modifications to an existing space. 'Refurbishment' tends to be utilised in relation to work that is cosmetic; typically, but not exclusively, decorative improvements and the upgrading of fixtures and fittings. However, it is the terminology of 'renovation' that sits more comfortably with the notion of an interior palimpsest as it refers to '*the act or process of repairing, renewing, or restoring to good condition*' as well as '*the act of reinvigorating or reviving*'[7] (dictionary.com, n.d.). The curative nature of restoration in association with the latter definition of revival, of rehabilitation, is of especial interest as this helps to justify the use of the term 'renovation' alongside 'adaption' (or adaptive reuse) throughout. How this reuse is achieved remains central to the notion of an interior palimpsest as typically there are limitless possibilities and design opportunities. The completion of these concise chapter summaries and clarification of terminologies allows this initial introductory chapter to return to an examination of the palimpsest's promotion of sustainability and narrative as two of the broad thematic themes of this book.

THE PALIMPSEST AND SUSTAINABILITY

The palimpsest's sustainable credentials have already been highlighted in relation to the environmental mantra of reduce, reuse, recycle, so contextually its re-examination is timely given our current preoccupation with these concerns. This stance, recognition even, would now benefit from further elucidation, as the circular economy is central to this ethical approach. Stahel is credited with developing a cyclical model that conceptually highlighted the need to move from a linear industrial economy that prioritises extraction and produce-use-discard, to a cyclical process linked to resource efficiency (1981)[8]. This approach acknowledges the impact of 'time' as this *'implies adaptability, flexibility, even humbleness – we do not know the future, but we can prepare for it'*[9] (Stahel preface to Baker-Brown, 2017: xiii). Temporality becomes a recurrent motif when considering the palimpsest as, since time immemorial, the circular economy was integral to our forebears' existence; from inherited properties and business premises, inter-generational living and coexistence as the norm, to the prevalence of 'spolia' from the Latin term for 'spoils' (essentially the reuse and relocation of architectural fragments and building materials within Antiquity) through to the inevitable hand me downs. As a sustainable philosophy, this is motivated as much by economic necessity as it is cultural and political gain. The move to an industrial mechanised society with a nomadic work force however heralded rapid change.

Baker-Brown's writings echo this stance, as for him a circular economy strives *'to turn our "throw-away, linear culture" into a "circular system" similar to the ecosystems found in the natural world'*[10] (2017: 01). This desire acknowledges humanity's impact globally captured by the term the 'Anthropocene Age', a term popularised by Crutzen (2002)[11] that defines a geological epoch that sadly recognises our negative impact upon both the climate and the environment and heralded a global aspiration to live in harmony with our planet.

Adaptive Reuse

This obvious desire for a more sustainable future contextualises the notion of the palimpsest and its relationship to adaption, the cultural legacy and the reuse of buildings. As both architecture and the design of interiors engage with an existing building or 'site', how this relationship is embodied and an approach advocated is, as ever, an interesting design conundrum. As already mentioned, prioritising the 'adaption' of existing buildings as a sustainable alternative to new build has to be an advantageous, ethically-driven starting point. Whilst an interior can just 'secrete' itself inside a building envelope, this form of architectural appropriation does little to acknowledge the value of an existing site, perhaps because the site has little architectural merit or history, is new, or because its cultural or architectural assets have simply been overlooked. In contrast, an adaptive reuse approach reimagines a new use for an existing building that deliberately acknowledges and responds to the inherent site condition, and its historical and architectural legacy. Rather than be demolished and end up as landfill, adaptive reuse projects typically celebrate the age, history and architectural inheritance of a building, whilst making it 'fit' for a new purpose and community of users. Central to this approach is an acknowledgment of a building's past life, a deliberate uncovering of site stories and architectural narratives that aims to celebrate the authenticity of the host building. By taking this approach buildings become celebrated as cultural artefacts, linked to a collective memory or history that remains implicit to the cultural value that bricks and mortar can embody as mnemonic aids. The notion of the palimpsest serves to reinforce the efficacy of a site via the imprint of memory and history, in essence its past life, incarnations or many 'interruptions', whilst encouraging building reuse through a process of rebirth.

Embodied Carbon versus Operational Carbon

Adaptive reuse as a methodology naturally embraces a detailed site analysis or site audit, an investigative appraisal that encompasses its history, assets and problems. This process encourages both remembrance and retention by identifying aspects of an existing building and its interior that are ripe for reuse. This approach, linked to a cyclical model of ethical production and consumption is central to both considering and limiting the 'embodied carbon' of any building. Put simply, embodied carbon is the carbon dioxide (CO_2) or greenhouse gas emissions (GHGs) associated with the material lifecycle of a building relating to its extraction, manufacture, transportation, installation and maintenance, as well as disposal. This is in contrast to 'operational carbon' (or the in-use phase of a building) that concentrates on energy consumption in relation to heating, cooling, ventilation and lighting. By appraising any existing building from the viewpoint of the palimpsest past incarnations become embedded into any future narrative, creating an interior that is full of preserved character; that links to the circular economy. Once existing buildings are reused, waste reduction occurs as the building shell, materials and surface treatments (in essence virgin materials) are potentially retained and embodied carbon reduced. Material durability, resilience and efficiency linked to circularity naturally embraces a patina of age, of wear and tear, of repair even that is responsive to a 'found' aesthetic. This trait encourages debate in relation to the perceived 'longevity' of a building, a stance that remains central to sustainable thinking as it challenges the notion of both a building or an interior with a limited shelf life locked into an endless unsustainable cycle of construction, deconstruction and reconstruction.

United Nations 17 Sustainable Developmental Goals

At the vanguard of this ambition is the United Nations 17 Sustainable Development Goals (UNSDGs) which aims to promote sustainable thinking globally by taking a holistic environmental, social and economic sustainability stance. Together they aim to tackle such issues as zero hunger, good health and well-being, gender equality and promote climate action by creating sustainable cities, communities and eco systems. Adopted by all member states of the United Nations in 2015, its reach is extensive and it aims to provide a *'shared blueprint for peace and prosperity for people and the planet, now and into the future'*[12] (United Nations, n.d.) that encourages responsible consumption and production. Recent industry-specific ambitions relate to the Royal Institute of British Architects (RIBA) and the British Institute of Interior Design (BIID) as both have produced publications that embrace the UNSDGs in relation to the built environment. RIBA's 'Sustainable Outcomes Guide' adopts eight of the UNSDGs (the nine exclusions for them relate to government policy and so fall outside of the remit of the construction industry). Their publication prioritises building reuse and the deep retrofitting of existing buildings as necessary measures towards achieving net zero operational and embodied carbon (Clark, 2019)[13]. The American Society of Interior Designers (ASID) has recently released a statement that focuses on three key tenets of design excellence, namely *'climate, health and equity'*[14] (n.d.), as a shared vision that provides guidance on sustainability. The European Parliament has also adopted a Circular Economy Package that sets legally-binding European Union (EU) targets for the recycling of waste and the reduction of landfill (2018)[15]. Khan's London Plan, developed as part of his mayoral election campaign, includes the ambition for London to be net zero carbon by 2030. An integral aspect of this is the desire for *'good growth – growth that is socially and economically inclusive and environmentally sustainable'*[16] (Greater London Authority, 2021: 26), and as such it prioritises the retention and reuse of existing built structures alongside repurposed and recycled materials (Greater London Authority, 2022)[17]. This all suggests that sustainability remains high on the agenda within the built environment from an organisational, governmental and intergovernmental level.

Sustainability Assessment Tools

To facilitate this approach assessment tools are available with BREEAM (Building Research Establishment Environmental Assessment Method) and LEED (Leadership in Energy and Environmental Design) being the two most commonly-used to measure and certify a building's performance globally. BREEAM's stated ambition is *'to mitigate the life-cycle impacts of a building on the environment'*[18] (2020: 4), whilst LEED promotes buildings that are safe, healthy, inclusive, smart, productive, efficient, equitable, sustainable, responsive and resilient (2023)[19]. For them *'a net zero carbon building is a highly efficient building that achieves a zero balance of carbon emissions emitted during operations'*[20] (LEED, n.d.).

This sustainable appraisal in relation to a palimpsestic approach to the built environment serves to highlight how this concept can encourage sustainable thinking through the integration of circularity, adaptive reuse and the limiting of embodied carbon. Consideration of any potential site or host building from the point of view of a palimpsest can encourage responsible consumption and production by embedding sustainability from the outset. This ethical stance embraces evidential and latent traces as a series of interior interruptions that naturally accommodates change, coexistence and encourages building and material reuse. The intention is to establish a palimpsestic reading of the built environment as a strategic exercise that enhances our understanding of both adaptive reuse and the interrupted interior, which helps to promote an ethically-informed sustainable strategy. This is a debate that is both topical and ongoing and remains of central importance to the study of architecture, interior architecture and interior design.

THE PALIMPSEST AND NARRATIVE

If sustainability is central to understanding an interior palimpsest, then the role of narrative has equivalence, as both environmental concerns and storytelling remain integral to this strategic approach. By rethinking building reuse as a series of perennial interruptions, questions remain regarding the role of narrative and its relationship to the palimpsest. Narrative's affiliation to design thinking and its role in weaving a story that enriches our occupation, aesthetic appreciation, interpretation and understanding of interior space will now be established. Luckily, there are many instances of the primacy of 'storytelling' within the built environment. Spiller believes that *'telling stories is one of humanities oldest pastimes'*[21] (2010: 128) and that it permeates all aspects of design, whilst the work and writings of Coates resonates with this approach, as he believes that storytelling gives meaning to the built world and that design should acknowledge and be routed in this expressive medium (2012)[22]. Many design and architectural practices would agree with this summation as they clearly embed narrative into the visceral experience and the intellectual property of their work. Interiors can translate the values of a specific brand within the retail sector as a staged encounter or story, whilst storytelling as a navigable route through museum and exhibition design is standard. Thematic interiors naturally embrace narrative as an active constituent of their design staging and scenography.

The Factual versus the Fictive

When describing narrative, stories, events and fictions remain conspicuous categorically, however there is an obvious point of tension or duality between the 'fictive' and the 'factually'-driven response. This is a recognition that suggests designers and architects can draw upon site narratives or place-centric stories, fabricated truths, supposition and proven historical events to frame their thinking. When considering narrative, the factual, historical or the anecdotal, the fictive, the reported or recounted reminiscence, even the myth, can all be referenced. A story can be analysed for how it is told and retold through a consideration of its structure or its sequential chronology, whilst textuality as a methodology that helps to frame meaning can be incorporated. Narrative can be utilised to depict, delineate, portray, expose or summarise a particular design or site approach. Storytelling can even be used to give a confessional or *in memoriam* feel to an interior. As the study of adaptive reuse is concerned with the reuse of existing building stock, uncovering forgotten site narratives that have become suppressed or overlain can inform this approach. This description stands in contrast to fictitious narratives that are developed through brand analysis or user experience in order to frame an interior atmosphere conceptually. This overview leads quickly to the supposition that storytelling in the built environment falls broadly into the following two categories:

1. **Design Fiction** – or fictive site responses composed of fictional site narratives
2. **Design Non-Fiction** – or factually-driven site responses composed of factual site narratives

All that remains is to appraise these two approaches and consider their relevance to the study of the palimpsest.

DESIGN FICTION OR FICTIONAL SITE NARRATIVES

If storytelling remains a vital ingredient of any design response, a detailed examination of the role of fictional narratives and their impact upon the interior is now valid. This analysis sets the scene and deliberately acts as a point of tension for a greater understanding of the factual narratives to follow. This appraisal aims to highlight how narrative is integral to and interwoven into a design response, whilst an understanding of both approaches gives the reader a holistic overview of storytelling in relation to the built environment. The purpose is to determine which narrative approach underpins the establishment of an interior palimpsest whilst simultaneously reinforcing a sustainable agenda. The three following precedents all proffer interesting insights in relation to fictional site narratives, their purpose and creation:

- A House for Essex by Grayson Perry and Charles Holland (FAT)
- 168 Upper Street by Amin Taha + Groupwork
- The Walled Off Hotel by Banksy

A House for Essex by Grayson Perry and Charles Holland (FAT)

Completed in 2014, 'A House for Essex', located in Great Britain, was an exciting collaboration between Charles Holland from the architectural collective FAT (Fashion, Architecture, Taste, now disbanded) and the artist and ceramicist Grayson Perry that explored an 'invented' narrative of design fiction. This creative invention justifies its inclusion as a prime example of a fictionally-driven site response. Luckily, both Perry and

Figure 1.2 and 1.3 A fictional character and her imagined life story creates a narrative of fabricated design fiction for this exuberant building and interior – A House for Essex, UK by Grayson Perry and Charles Holland (FAT).
Copyright: Jack Hobhouse.

Holland share certain preconceptions relating to class, cultural stereotypes, taste and meta-narratives resulting in a common language that informed the project. Taking inspiration from the cultural preconceptions linked to the inhabitants of the region, Holland and Perry successfully embedded a fictional narrative into this exquisitely crafted object. Built as a holiday rental for Living Architecture (an organisation spearheaded by Alain de Botton that commissions residential buildings to promote and create dialogue around contemporary architecture), the project was about 'Essex', a region with strong connections for both protagonists. Perry explains how this location is inextricably linked to cultural class stereotypes: '*Essex was sort of Home Counties but not the nice stockbroker-y bit. It was mainly Cockney overspill. That's the real character of Essex: it's where the Cockneys meet yokels*'[23] (Perry cited in Rose: 2014). Taking inspiration from this rich cultural landscape, the fictional resident of the house, Essex everywoman 'Julie Cope' became a literal conglomeration of these class-based observations. Perry developed this fictional biography of class mobility, from simple working-class origins to middle-class aspirations via a poem entitled 'The Ballad of Julie Cope' and this inspired the design of both the building and its interior:

> Julie was born on Canvey Island in the great floods of 1953. She was a flowerchild who, after a failed marriage, had a 'redemptive second act' in Colchester with her second husband, Rob, an IT consultant. Unfortunately, last year she was hit by a curry deliveryman riding a moped and killed. The grieving Rob built this house near Wrabness as a memorial to her[24].
>
> (Turner, 2015: 80)

Informed by folk, vernacular architecture and populist buildings such as wayside chapels, places of pilgrimage, churches and follies, a design soon began to emerge that exemplified the notion of 'Gesamtkunstwerk'; a cohesive vision or 'total work of art'. Intended as a secular shrine dedicated to Julie's fictional life, this '*constructed reality*'[25] (Pritchard, 2015: 39) biographically celebrates her life, marriage and death. Julie remains omnipresent, her iconography enshrined within the architecture, ceramics, soft furnishings, furniture, fixtures and fittings, which informs the decorative excess both internally and externally. This example beautifully highlights how fictional site narratives can invest a work of architecture and its interior with an infectious exuberance driven by 'imagined' storytelling, in this instance inspired by a specific locality and its inhabitants.

168 Upper Street by Amin Taha + Groupwork

In contrast, 168 Upper Street, located in London and designed by Amin Taha and Groupwork, is a delightful example of deliberate '*narrative* misremembering'[26] (Taha cited in Mollard, 2017: 24) that is 'inspired' by its context rather than completely defined by it. The property occupies the demolished end section of a Victorian terrace, a former World War II bombsite that eschews the faithful re-creation of this lost nineteenth-century building. Instead, it chooses to 'play' with memory, of the imaginary versus the factual, and this juxtaposition informs a process that acknowledges the practice of artist Rachel Whiteread and her 'casting' of overlooked, forgotten and negative spaces. Groupwork's design is the result of extensive contextual and historical exploration, digital mapping, modelling and drawing of the lost building's twin still standing proudly at the end of the row. This is an exploration that forms the basis for their monolithic, concrete, terracotta hued cast architectural creation that is: *'Painstakingly cast into machine milled polystyrene moulds constructed from photographs and a digital 3D survey of the mirror corner at the other end of the terrace, it creates, rather than re-creates . . .'*[27] (RIBA, 2019).

The emphasis on 'creation' rather than 're-creation' is significant as Groupwork's process of reimagined rather than reconstructed reality deliberately subverts the existing site narrative. Rather than producing a factually-accurate response, or simple re-creation, it creates a new fiction for the building (albeit one that is inextricably wedded to its location). The deliberate inclusion of architectural errors and mistakes is integral to the process and this takes a variety of forms. From happy accidents on site where pieces of the formwork relating to the casting were displaced (resulting in a section of the roof parapet being left blank and reappearing elsewhere), to complex digital drawings of the façade full of architectural ornamentation that when cast became broken during the construction stage, all were retained. These 'digital' and 'build' inaccuracies relate to a rethinking of the past as 'imperfectly' remembered and were incorporated as integral to the new site narrative. Original windows became cast low relief infilled blank imprints accommodating new contemporary openings that punched through the façade (as dictated by the internal planning and function). This deliberate disruption of the neo-classical street façade, alongside the retained 'inaccuracies', transforms their response from an act of simple mimicry, or a faithful facsimile, to an elaborate fictional subterfuge. Instead, the *'the errors reinforce the theme that the memories from which we reconstruct are as flawed and selective as the narratives which we attempt to project from the past'*[28] (Coe cited in Ravenscroft, 2019). Completed in 2017, this architectural response tells a unique story, but a story that looks to the past for inspiration and then 'subverts' that knowledge as the new fictional narrative supersedes reality.

Figure 1.4 An artistic response creates a fictionalised interior narrative inspired as a commentary on the ongoing regional conflict – The Walled Off Hotel, Bethlehem, Palestine by Banksy.
Source: YellowFratello, CC0, Wikimedia Commons. Licensed under the Creative Commons Attribution-Share Alike 3.0 Unported license: https://creativecommons.org/licenses/by-sa/3.0/deed.en.

The Walled Off Hotel, Banksy

Finally, graffiti artist Banksy's celebrated international foray into interior design via 'The Walled Off Hotel' is another useful example of a fictional site narrative that relies on design fiction. Located in the Palestinian-controlled West Bank territory of Bethlehem, this refurbishment deliberately acknowledges regional political tensions as the site exists in close proximity to the wall that divides Israel from the Palestinian-controlled territories. Banksy's oeuvre showcases an artist who is never shy of making political, social or cultural statements (often with his trademark irreverent sense of humour) and this project is no exception. Whilst acknowledging its context, the interior (originally a pottery workshop) '*has a dystopian colonial theme, a nod to Britain's role in the region's history, the reception and tea-room a disconcerting take on a gentleman's club where a self-playing piano provides an eerie soundtrack*'[29] (Graham-Harrison: 2017). Conceived as a hotel, museum and gallery, the project aims to highlight the ongoing conflict and provide a space for Palestinian artists. Boasting the 'worst hotel view in the world', the interior and the art it contains is politicised to reference the ongoing regional conflict. Tear gas canisters wreathe object d'art in plumes of smoke, surveillance cameras abound, graffiti cherubs hover above the automated grand piano replete with oxygen masks, whilst a carefully-positioned telescope capitalises on the non-view. Every accessory and decorative flourish acknowledges the conflict in the creation of a cohesive interior narrative; the design of the interior is clearly a work of fiction that is 'inspired' by its immediate location and the long-running conflict.

Conclusions

In conclusion, these three examples have explored fictional site narratives as works of imagined design fiction. These precedents all acknowledge the power of storytelling to inform the creative direction, project realisation and the experiential quality of an interior. Perry and Holland's a House for Essex utilises the 'fictional' biography of a character linked to regional cultural stereotypes in both its inception and creation. Groupwork's reimagining of a lost building acknowledges architectural antecedent but 'disrupts' this with deliberate creative errors, allowing the new site narrative to misremember the past in the creation of a new fictitious reality. Finally, Banksy's Walled Off Hotel (the name surely a deliberate pun on the global hotelier brand The Waldorf Hotel) utilises the conflict to create a politicised interior narrative that aims to accentuate the regional tensions by encouraging debate and discussion. Usefully, this appraisal highlights the integral role of narrative in the design of the interior but acknowledges how in this instance creative licence 'subverts' storytelling. The inclusion of two interiors reimagined by artists is deliberate as this approach often aligns with artistic practice. Crucially, fictionalised storytelling does not reinforce the notion of an interior palimpsest as it pays less attention to the 'reality' of the site; instead, it uses its context as a creative springboard in the development of a suitable design response. Similarly, the absence of a sustainable agenda is another important omission. In summation, fictitious site responses relating to imagined design fictions tend to overlay, embellish or even invent narratives; they are stories that are imposed contextually. This form of storytelling tends to have a looser grasp on the historical reality of the site, but is often sensitive to broader themes relating to the wider regional, political or cultural context. Typically, they are 'inspired' by their context rather than completely defined by them.

DESIGN NON-FICTION OR FACTUAL SITE NARRATIVES

In contrast, factually-driven site responses or design non-fiction tends to colonise adaptive reuse or more architectural interpretations of the interior as they pay homage to the history of a site and the original or 'host' building that retains and celebrates historical site narratives. Littlefield and Lewis's seminal publication *Architectural Voices* is a strong advocate for the advantages of 'listening to old buildings' (2007)[30] by teasing out the narratives that they inevitably contain. Similarly, the writings of Hollis revel in their descriptions of detailed historical interior vignettes and the stories they embody as repositories or 'memory palaces' (2013)[31], whilst the French philosopher Perec documented through his writings an obsessive interest in narrative space. His observational interpretation of people's everyday lives was concerned with stories linked to specific locations. His *'infra-ordinary'* (which stood in direct contrast to the extraordinary) was about the diurnal patterns found in the mundane, the ordinary and the repetitive (2008[1974]: 210)[32]. The writings of German philosopher Benjamin conclude this literary appraisal of factually-driven site responses by identifying the '*traces*' left by an interior's inhabitant and their significance in forming an ongoing temporal story of occupation (2002[1982]: 20)[33].

This suggests that place-centric narratives are 'defined' by their physical context and represent the 'life' of a building, but additionally these stories are informed socially by the inhabitants that live within its confines. Luckily, there are many examples to draw upon in the exploration of 'factual' site narratives. The forthcoming selection highlights a veritable smorgasbord of approaches that combines storytelling and site responsiveness as central constituents of design within the built environment. In contrast to the fictional site narratives already examined, this section explores in detail factual site narratives or design non-fiction and sets out to establish their pivotal relationship to both adaption and the interior palimpsest. Once again, an analysis of three carefully-selected precedents offers insights into site understanding and response, and these examples are:

- Crystal Houses by MVRDV
- Ryú Peel by Ménard Dworkind Architecture & Design
- Croft Lodge Studio by Kate Darby Architects and David Connor

Crystal Houses by MVRDV

The first precedent for consideration is MVRDV's re-creation of two traditional Flemish townhouses that justifies its inclusion because of its prioritisation of the past, as a revelatory, if factually-driven site story of design non-fiction. Located in Amsterdam's Museum Quarter, the design response sought to address the increased globalisation and homogeneity in relation to high-end retail streetscapes. PC Hooftstraat was initially a nineteenth century residential street created by the reknowned architect Pierre Cuypers. Previous developments of the shop facades had resulted in a loss of the vernacular character at street level that did little to enhance the historical remnants above. Dutch-based architectural practice MVRDV were asked by their client Warenar (an Amsterdam-based company who focus on prestigious retail real estate) to consider this issue. Their solution was to recreate in 'glass' the two original historical townhouses as replacements to the modified street elevation. The result is a crystalline façade that aims to enhance the architectural quality of the street, but achieved in a pellucid manner: *'the near full-glass façade mimics the original design, down to the layering of the bricks and the details of the window frames'*[34] (MVRDV: n.d.). This faithful re-creation in all but material allowed for greater transparency, essential for a retail space occupied by global brands, whilst creating an iconic flagship store that was sensitive to the brutalised historic streetscape. MVRDV's extensive drawing documentation captured the existing site condition, in essence what was to be demolished (the buildings street facade plus the contemporary ground floor retail alterations) alongside drawings of the original, historic façade (prior to any alterations). This became the 'template' for the new glass façade that was then stretched vertically to 'fit' on the newly rebuilt and slightly larger building volume.

This poetic approach uncovered a factual narrative via historical research, place-centric storytelling in the development of a solution that is 'defined' by its physical context and the previous 'life' of a building, but contemporises this via the use of a seductive new material. This strategic response resonates with the approach taken by Groupwork in the previous examination of design fictions, but critically differs because it relies on historical accuracy rather than a fictionalised misremembering. This place-centric revelatory story also has sustainable credentials (surprising given the fact that the original was demolished) as the intention was to create a blueprint that could be emulated globally, reinforcing regional identity within historic city or town centres. The development of glass bricks in the near full-glass facade ensured that future waste materials were minimised, as all the glass components could be recycled.

Ryú Peel by Ménard Dworkind Architecture & Design

Following on from this initial analysis, this second precedent shares a creative process that acknowledges its site history, but achieves this via a revelatory technique that embraces a factual narrative or design non-fiction. Ryú Peel as a brand are a chain of Japanese Sushi restaurants and the interior of their Montreal-based restaurant, designed by Canadian practice Ménard Dworkind Architecture & Design (MRDK), embraced an approach that celebrated place-centric storytelling and capitalised on the host building's rich decorative history. A fortuitous discovery whilst undertaking a survey on site revealed the walls had been covered by millwork, with their removal disclosing the original wall surfaces as a rich 'tapestry' that documented the tastes of previous occupants. This revelation led to a design conundrum as to whether this decorative heritage should be discarded or concealed, or be integrated and celebrated within the interior refurbishment. MRDK decided to retain the old to inform the new, as they remarked: *'instead of treating the space as a blank canvas, (we) embraced the existing building fabric in all its layered history and*

Figure 1.5 and 1.6 This interior is responsive to a factual site narrative via an accumulation of decorative detritus; this sustainable retention is integral to the ongoing interior story – Ryú Peel, Montreal, Canada by Ménard Dworkind Architecture & Design (MRDK).

Copyright: Ménard Dworkind Architecture & Design.

imagined imperfections'[35] (MRDK cited in Yatzer, 2019). By integrating the Japanese philosophy of Wabi Sabi (wholly appropriate given that this is a restaurant that celebrates Japanese cuisine), the idea of an aged interior dating from 1840, of the heritage of a space, of the celebration of imperfections in the quest for authenticity was justified. Decorative temporality in this instance became a conscious design choice:

> Whereas most architects would have excised all traces of past occupants, in this case, the building's walls have been turned into an in-situ archaeological display where peeling wallpaper, decaying plasterwork, exposed concrete and hand-drawings comprise a visual palimpsest whose beauty lies in its myriad of imperfections[36].
>
> (Yatzer, 2019)

In contrast to this urban decorative decay, a new suspended bespoke timber structure incorporates lighting and utilises traditional Japanese craft techniques juxtaposed against the patina of retained decorative dereliction prevalent within the restaurant interior. New contemporary bar fixtures and furniture serve to enhance this narrative of temporal duality further. The justification for this example's inclusion is now apparent via its embracing of design non-fiction, in its capturing of a factual narrative based upon the building's previous occupation and decorative heritage. This place-centric revelatory story clearly has sustainable credentials as it prioritises the reuse of virgin materials alongside its consideration of the continued life of the building.

Croft Lodge Studio by Kate Darby Architects and David Connor Design

Finally, the inclusion of Croft Lodge Studio by Kate Darby Architects and David Connor Design brings to a natural conclusion this analysis of factually-driven site narratives. This precedent's incorporation is evident once the creative approach strategically is considered, essentially a form of '*extreme conservation*'[37] (Spicer: n.d.). Located on Bircher Common, Herefordshire and near Croft Castle on land owned by the National Trust, the site, a ruined seventeenth-century cottage and stables, was in such a poor state that its very dereliction became the catalyst for refurbishment. Early discussions with conservation officers emphasised the need to 'conserve' the curtilage listed ruins, so, if the site could not be cleared and built on anew any approach needed to embrace its current ruination. The solution was to preserve everything by encasing it within a new contemporary

Figure 1.7 A factual site narrative celebrates a poetic ruin by preserving it within a new building in the creation of an extraordinary synergy of the old and the new – Croft Lodge Studio, UK by Kate Darby Architects and David Connor Design.
Copyright: James Morris.

structure, an approach that was also driven by financial necessity. A new steel frame sits over and envelopes the ruin, whilst black corrugated external cladding (a nod to the many agricultural buildings found in the region and linked to a keen interest in rural architecture and place-based materiality) creates the necessary enclosure. This new structure's purpose is two-fold; firstly it brings the ruin up to modern day specifications relating to protection, warmth and energy efficiency, whilst secondly, and most importantly, it stabilises the ruin. As Conner explains:

Figure 1.8 The plans highlight the remnants of the original seventeenth-century ruin and the new contemporary 'ghost' structure enveloping the preserved remains.
Copyright: Kate Darby Architects and David Connor Design.

Figure 1.9 Image of the ruin before building work commenced – Croft Lodge Studio, UK by Kate Darby Architects and David Connor Design.
Copyright: Kate Darby Architects and David Connor.

In many ways it was an incredibly pragmatic job. We've got an old building; we can't knock it down. We didn't really want to repair it; it was beyond repair so what do you do? You stick something over it, you put another building on top of it to protect it[38].

(Dirksen, 2019)

This process of protective encasement was responsive to the rebuild of the collapsed stable building, whilst a new contemporary, neutral volume recreates the 'ghost' of what was once there. By echoing the original building's footprint the new volume protects the architectural remnants, whilst the conservation approach of *'not touching anything'*[39] (Dirksen, 2019) has resulted in an elaborate synergy of the old and the new that exploits a narrative of continued occupation. This simple decision created a form of architectural storytelling that wholeheartedly embraced the existing site narrative, celebrating the weathering and patina of age as something that cannot be easily replicated.

The result is a frozen moment in time, an architectural snapshot of an aged, rotten and worm-eaten timber ruin replete with ivy, cobwebs, a bird's nest, blown lathe and plaster, found objects and dilapidated, unglazed metal window frames. Even the dust becomes celebrated, almost 'fetishised' against a new neutral contemporary backdrop of whitewashed walls and ceilings. The ruin becomes a museum piece, a work of art, a 'found' object that is preserved for posterity 'framed' by the new building and it is this temporal juxtaposition that makes the project so appealing. Operating as a studio with accommodation that can be converted into a dwelling as needed, the completed project has been included because it captures a factual site narrative based upon a true palimpsest, a dialogue that looks simultaneously to the past and the present. The resultant interior revels in its celebration of architectural poetics, of vestigial remnants that inform both the strategy of conservation and an interior atmosphere of decay and degradation. This approach ensures the ruin has a continual life whilst the building timeline is clearly legible via both preservation and a contemporaneous addition.

Conclusion

To conclude, these three precedents have all highlighted how factual narratives and an interest in design non-fiction harnesses the power of place-centric storytelling to inform the creative direction, realisation and experience of an interior. Crystal Houses by MVRDV documents both the existing and the historical condition in the quest for a narrative template that reinstates the original building façade, but revitalises it with the use of a new material. This ghostly evocation of the building's inception stays true to its original incarnation that now has an enhanced regional identity that exploits its new materiality. In contrast, Ryú Peel by Menard Dworkind Architecture & Design embraces a found aesthetic, utilising a site audit to document the building's life. This is an approach that prioritises the retention of decorative traces as an archaeological narrative of occupation consciously complemented by the new. Similarly, Croft Lodge Studio by Kate Darby Architects and David Connor Design is responsive to the limitations of the listed, derelict site by preserving in situ the cottage ruins. This results in a narrative of both resurrection and dilapidation, of a 'found' contextual theatricality offset by the deliberate unobtrusiveness of the new. This succinct appraisal aims to highlight the integral role of factual site narratives to the establishment of an interior palimpsest. This is a site narrative that engenders a sustainable dialogue of cohabitation that looks to the past for inspiration, prioritises retention, but always looks to modify or incorporate the new.

CHAPTER CONCLUSIONS

If a palimpsest is interested in narrative and encourages sustainability by retaining temporal collisions between the past and the present, the factual site narratives already discussed become a useful gauge for both engendering and reinforcing this strategic approach. For the interrupted interior, palimpsestic storytelling becomes a process that is investigative, inquisitive, even archaeological in nature. How to tell a story is always a matter of creative interpretation, but a palimpsestic approach clearly celebrates its heritage, consciously incorporating factual site narratives, its site reality, by embracing design non-fiction. An acknowledgment of the host building's life cycle in any reformation however becomes tempered by the desire to extend or enhance the existing site narrative. This temporal layering consciously embraces a revised meaning through a continual process of writing, rewriting and even unwriting. The use of and inclusion of these textual metaphors in relation to the establishment of an interior palimpsest is deliberate and will be discussed in more detail in the forthcoming chapters.

Finally, factual narratives clearly have a stronger synergy with sustainable concerns in their embracing of adaptive reuse and consideration of a host building's longevity, history and circularity. An ethical approach that values authenticity and retains the original, and uses it to forge a new contemporary alliance of change, instilled with the essence of its biographical past. By actively 'listening' to its context, creative inspiration becomes situational. A palimpsestic approach when related to the built environment clearly prioritises factually-driven site storytelling and encourages sustainable thinking. Chapter Two will now explore in detail the character and interdisciplinary nature of the palimpsest utilising a comparative analysis to develop a definitive interpretation in relation to its manifestation within the interior realm.

REFERENCE LIST

1 Clark, G. 2019. 'RIBA Sustainable Outcomes Guide'. *RIBA*. [online pdf,v3]. Available at: https://www.architecture.com/knowledge-and-resources/resources-landing-page/sustainable-outcomes-guide [accessed 16 June 2022].
2 Clark, G. 2019. 'RIBA Sustainable Outcomes Guide'. *RIBA*. [online pdf,v3]. Available at: https://www.architecture.com/knowledge-and-resources/resources-landing-page/sustainable-outcomes-guide [accessed 16 June 2022].
3 BIID, 2021. 'BIID Sustainable Specifying Guide'. *BIID*. [online pdf]. Available at: https://biid.org.uk/resources/sustainable-specifying-guide [accessed 16 June 2022].
4 Norberg-Schulz, C. 1984. *Genius Loci: Towards a Phenomenology of Architecture*. (1980). New York: Rizzoli.
5 Dictionary.com. n.d. 'Narrative definition'. *Dictionary.com* [online]. Available at: https://www.dictionary.com/browse/narrative [accessed 1 June 2022].

6 Stevenson, A. and Waite, M. 2011. *Concise Oxford English Dictionary*. Oxford: Oxford Dictionary Press, 12th edition.

7 Dictionary.com. n.d. 'Renovation definition'. *Dictionary.com* [online]. Available at: https://www.dictionary.com/browse/renovation [accessed 31 august 2023].

8 Stahel, W.R. (1981) 'The Product-Life Factor' [online pdf]. *Google Scholar*. Available at: https://p2infohouse.org/ref/33/32217.pdf [accessed 22 April 2023].

9 Baker-Brown, D. 2017. *The Re-Use Atlas: A Designer's Guide towards a Circular Economy*. London: RIBA Publishing.

10 Baker-Brown, D. 2017. *The Re-Use Atlas: A Designer's Guide towards a Circular Economy*. London: RIBA Publishing.

11 Crutzen, P.J. (2002). 'Geology of Mankind', *Nature*, 415: 23.

12 United Nations. n.d. 'Sustainable Development Goals'. *SDGS* [online]. Available at: https://sdgs.un.org/goals [accessed 16 June 2022].

13 Clark, G. 2019. RIBA Sustainable Outcomes Guide. *RIBA*. [online pdf]. Available at: https://www.architecture.com/knowledge-and-resources/resources-landing-page/sustainable-outcomes-guide [accessed 16 June 2022].

14 American Institute Of Interior Designers (ASID). n.d. 'ASID's Position Statement on Climate, health, and Equity'. *ASID.ORG* [online]. Available at: https://www.asid.org/position-statements [accessed 14 January 2023].

15 European Parliament (EU). 2018. 'Circular Economy Package: Four Legislative Proposals in Waster'. *European Parliament*. [online pdf]. Available at: https://www.europarl.europa.eu/RegData/etudes/BRIE/2018/614766/EPRS_BRI(2018)614766_EN.pdf [accessed 7 January 2023].

16 Greater London Authority. 2021. 'The London Plan'. *London.gov.uk* [online]. Available at: https://www.london.gov.uk/sites/default/files/the_london_plan_2021.pdf [accessed 19 October 2022].

17 Greater London Authority. 2022. 'Whole Life-Cycle Carbon Assessments'. *London.gov.uk* [online]. Available at: https://www.london.gov.uk/sites/default/files/lpg_-_wlca_guidance.pdf [accessed 19 October 2022].

18 BREEAM. 2020. 'BREEAM UK Refurbishment and Fit-out 2014, Non-domestic Buildings'. *BREEAM*. [online technical manual SD216 2.0]. Available at: https://files.bregroup.com/breeam/technicalmanuals/ndrefurb2014manual/ [accessed 11 April 2023].

19 LEED. 2023. 'About LEED: Downloadable Presentation' [online presentation]. LEED. Available at: https://www.slideshare.net/USGBC/about-leed-downloadable-presentation [accessed 22 April 2023].

20 LEED. n.d. 'Net Zero' Available at: https://www.usgbc.org/about/priorities/net-zero [accessed 22 April 2023].

21 Spiller, N. 2010. 'Telling Stories'. *Architectural Design*, Jan/Feb, 203 (80): 128–9.

22 Coates, N. 2012. *A.D. Primers: Narrative Architecture*. Chichester: John Wiley & Sons Ltd.

23 Rose, S. 2014. 'The Making of a House for Essex'. *Living-Architecture* [online]. Available at: https://www.living-architecture.co.uk/the-houses/a-house-for-essex/architecture/ [accessed 16 June 2022].

24 Turner, C. 2015. 'A House for Essex'. *Icon*, 145: 78–85.

25 Pritchard, O. 2015. 'Study: A House for Essex, by FAT and Grayson Perry' (Appraisal 2). *Architect's Journal*, 242 (7–8): 34–49.

26 Mollard, M. 2017. 'Making an Impression'. *The Architectural Review*, 242 (1444): 18–25.
27 RIBA. 2019. '168 Upper Street'. *RIBA* [online]. Available at: https://www.architecture.com/awards-and-competitions-landing-page/awards/riba-regional-awards/riba-london-award-winners/2019/168-upper-street [accessed 16 June 2022].
28 Ravenscroft, T. 2019. 'Amina Taha creates distorted replica of 19th-century London terrace block'. *Dezeen* [online]. Available at: https://www.dezeen.com/2019/04/01/amin-taha-groupwork-168-upper-street-london/ [accessed 20 April 2022].
29 Graham-Harrison, E. 2017. 'Worst View in the World: Banksy opens hotel overlooking Bethlehem Wall'. *The Guardian 3 March* [online]. Available at: https://www.theguardian.com/world/2017/mar/03/banksy-opens-bethlehem-barrier-wall-hotel [accessed 16 June 2022].
30 Littlefield, D. and Lewis, S. 2007. *Architectural Voices: Listening to Old Buildings*. Chichester: Wiley-Academy.
31 Hollis, E. 2013. *The Memory Palace: A Book of Lost Interiors*. London: Portobello Books.
32 Perec, G. 2008. *Species of Spaces and Other Pieces* (1974). Edited and translated by John Sturrock. London: Penguin Classics.
33 Benjamin, W. 2002. *The Arcades Project*. Paperback edn. (1982). Translated by Howard Eiland and Kevin MaLaughlin, based upon the German volume edited by Rolf Tiedmann. Cambridge, Massachusetts and London: The Belknap Press of Harvard University Press.
34 MVRDV. n.d. 'Crystal Houses'. *MVRDV* [online]. Available at: https://www.mvrdv.nl/projects/240/crystal-houses [accessed 16 June 2022].
35 Yatzer. 2019. 'RYU Peel: A Sushi Bar in Montreal embraces the Aesthetic Philosophy of Wabi-Sabi'. *Yatzer* [online]. Available at: https://www.yatzer.com/ryu-peel [accessed 16 June 2022].
36 Yatzer. 2019. 'RYU Peel: A Sushi Bar in Montreal embraces the Aesthetic Philosophy of Wabi-Sabi'. *Yatzer* [online]. Available at: https://www.yatzer.com/ryu-peel [accessed 16 June 2022].
37 Spicer, A. n.d. *Extreme Conservation: Croft Lodge Studio* [online video]. Available at: https://www.andrewspicer.co.uk/extreme-conservation-croft-lodge-studio [accessed 16 June 2022].
38 Dirksen, K. 2019. 'Medieval Cottage encased within new home embraces layers'. *Fair Companies* [online video]. Available at: https://faircompanies.com/videos/transient-18th-c-cabin-inside-new-home-embraces-time-layers/ [accessed 16 June 2022].
39 Dirksen, K. 2019. 'Medieval Cottage encased within new home embraces layers'. *Fair Companies* [online video]. Available at: https://faircompanies.com/videos/transient-18th-c-cabin-inside-new-home-embraces-time-layers/ [accessed 16 June 2022].

What is a Palimpsest? Two

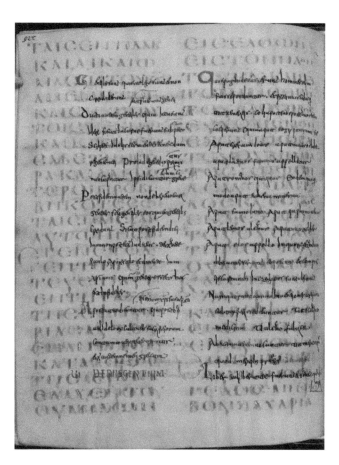

Figure 2.1 This image is a classic example of a palimpsest; writing material (such as a parchment or a tablet) which is reused but imperfectly cleaned, allowing writings from different historical periods to coexist. The Codex Guelferbytanus 64 Weissenburgensis contains text from sixth and the thirteenth century.

Source: Codex Guelferbytanus 64 Weissenburgensis, Wikimedia Commons. Licensed under the Creative Commons Attribution-Share Alike 3.0 Unported license: https://creativecommons.org/licenses/by-sa/3.0/deed.en.

WHAT IS A PALIMPSEST?

Whilst Chapter One set the scene and introduced broad palimpsestic themes relating to interior interruptions, narrative and sustainability, this chapter's focus differs as it sets out to proffer a definitive definition of a palimpsest. An integral aspect of this investigation will be the establishment of the palimpsest's persuasive reach via an interdisciplinary analysis. It is an inquiry that highlights its affinity to the creative industries via an exploration of the artistic, literary, cinematic, design and architectural realms. This discussion, by creating a series of comparative parameters, culminates in the establishment of the typical traits and preoccupations of an interior palimpsest. The forthcoming investigation facilitates a theoretical debate framed by the question: **what is a palimpsest?** This inaugural question becomes a useful starting point that commences with an etymological enquiry which begins with a dictionary definition, included here for clarity:

palimpsest /pæləmpˌsɛst/ **noun**

1 technical: a very old document on which the original writing has been erased and replaced with new writing

2 formal: something that has changed over time and shows evidence of that change[1].

(Britannica, 2022)

This clarification relates to the preconceptions that many already have in relation to a palimpsest, a medieval document often written on parchment or velum that, because of the scarcity of writing materials, has been recycled. The word palimpsest has been developed from the Greek '*palimpsēstos*' with '*palin*' relating to '*again*' and '*psēstos*' to '*scraped, rubbed*'[2] (dictionary.com, n.d.), and this notion of reuse remains critical to any burgeoning definition. By delving into the world of Antiquity, both the *Encyclopaedia of Ancient Literature* (Cook, 2014)[3] and the *Encyclopaedia of Ancient Christianity* (Milazzo, 2014)[4] reinforce this preoccupation by proffering a definition that includes a writing surface such as a tablet or manuscript that has been reused. This fixation is important as it helps to explain '*codex rescriptus*', a term that is generally preferred today to palimpsest and is a literal translation from the Greek '*I scrape afresh*'[5] (Milazzo, 2014). Interestingly, this form of erasure (with earlier writings being washed or scraped off) was often imperfect, as '*traces of the previous writing would resurface*'[6] (Milazzo, 2014). During the Middle Ages both the cost and scarcity of materials led to ancient manuscripts (or codices to use the correct term) that were damaged or whose content was regarded as disposable (think out of date, deemed heretical or simply unintelligible) being recycled. A classic example is located in the Biblioteca Medicea Laurenziana, Florence and this is a palimpsest containing three important layers:

> The most ancient layer of which dates back to the 9th c. and contains Theodoret of Cyrrhus's 'Historia philothea'; the mid-layer dates to the 11th c. and contains Homer's Iliad; the most recent layer dates to the 13th c. and contains various texts among which are works of Aristotle[7].
> (Milazzo, 2014)

This historical layering allowed lost texts ('*scriptio inferior*' or the faint remnants of 'underwriting') to be recovered or to re-emerge alongside the current text ('*scriptio superior*' or 'overwriting'). Globally-significant examples of palimpsests include the 'Codex Ephraemi Rescriptus' currently located in France that famously includes portions of the Old and New Testaments written in Greek from the fifth century, overwritten by the works of Ephraem the Syrian from the twelfth century; whilst the Archimedes Palimpsest captures seven treatises written by the Ancient Greek mathematician, within a medieval prayer book from the thirteenth century. The Biblioteca Capitolare, or the library of the Chapter of Verona (originally founded as a scriptorium), can lay claim to the most famous palimpsest. Discovered in the nineteenth century this palimpsest contains an almost complete text of the Institutes of Gaius (the only surviving textbook of Classical Roman Law), underwritten beneath the fifth-century theological writings of Saint Jerome.

This notion of 'reuse' remains central to any palimpsestic definition and is why it resonates in today's environmentally-conscious era (as already discussed in Chapter One), as a true palimpsest occupies a temporal zone that simultaneously looks to the past and the present, that secures the future. This temporal relationship embraces both erasure and alteration, whilst celebrating vestigial traces and remnants. The importance of memory, layers and imperfections become recurring traits that initially relate to ancient documents, but have a familiar resonance with the built environment. Evidential change in the form of **under-** and **overwriting**, even **unwriting** (useful terms because they reference the palimpsestic process), is central to the establishment of a relevant, if incipient, terminology. So, if the palimpsest is primarily concerned with history and layering, it prioritises multiple layers of under- and overwriting as mnemonic traces, a remnant of old, older, new and newer. After all, as Hollis writes on the historiography of interiors: '. . . *all interiors, are, to some degree or another, made out of remnants of others*'[8] (2010: 105).

PALIMPSESTS: TYPES, DIFFERENCES AND COMMONALITIES

Discussions of the palimpsest's inherent interdisciplinarity can be found to inform the fields of archaeology and landscape, perhaps because practitioners from these disciplines are dealing with strata in their investigation of historical sites. The physical act of the archaeological 'dig' aims to reveal a timeline of occupation and habitation through accumulated 'traces'. Similarly, it has relevance to architecture and design via an understanding of the ongoing biography of a building in relation to its reuse. Film, art and, more recently, linguistics and literary theory have all fallen under the persuasive spell of the palimpsest. By examining various creative disciplines, namely the artistic, literary, cinematic and architectural, this preoccupation will be analysed before returning cyclically to the central premise of interior interruptions.

By investigating palimpsestic thinking across multiple creative disciplines, the intention was not to provide an exhaustive linear chronology (the length of this book and the focus on the interior prohibits this). Instead, it was to select, discuss and highlight the palimpsest's pertinence, reach and influence across a wide range of disciplines. This approach, essentially a comparative analysis, aims to establish important recurring concerns, culminating with the establishment of the characteristics and traits that typically inform an interior palimpsest. This selection is by nature expansive, allowing for connections and comparisons across multiple approaches, theories, precedents and individuals, and is an analysis that precipitates a delving into and complete immersion in the realm of the palimpsest.

The Art Palimpsest

When considering the world of art, the palimpsest has an obvious equivalence with the artistic practice of '*pentimento*', a term taken from the Italian '*pentirsi*' meaning 'to repent', and the preoccupations previously outlined. Earlier drawn revisions or alterations, alongside the practice of overpainting, become apparent on the surface of a canvas as it ages and the paint thins. These 'repentances' appear as traces, ghostly outlines that were initially

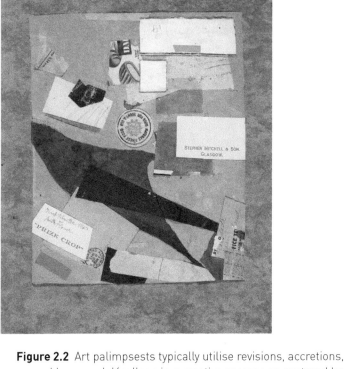

Figure 2.2 Art palimpsests typically utilise revisions, accretions, assemblages and décollage in a creative process as captured by the art of Schwitters.
Source: Kurt Schwitters, Wikimedia Commons. Licensed under the Creative Commons Attribution-Share Alike 3.0 Unported license: https://creativecommons.org/licenses/by-sa/3.0/deed.en.

overpainted, erased by the artist, but have resurfaced to inform the palimpsestic story of a particular work of art. One of the most famous artistic examples is the 'Old Guitarist' by Picasso, as the painting contains the ghostly apparition of a woman. Rembrandt's 'Flora' from the seventeenth century is another classic example that famously contains a double hat brim as a revenant of an earlier composition. This preoccupation with layering has obvious similarities with the artistic practice of collage and décollage, as both are examples of palimpsestic art given their associated techniques exploit the surface condition through a process of addition and subtraction. Schwitters famously employed collage as a method of artistic expression by collecting the discarded detritus and ephemera of everyday life to form 'Merz Pictures'[9] (Tate, n.d.). These assemblages utilised found objects such as train tickets, newspapers, used envelopes, even discarded cigarettes to form new ripped, assembled and overlapping two-dimensional compositions. Over time this methodology naturally increased in scale, eventually forming immersive environments such as the 'Merzbau', a transformative process that infected the artist's family home and was a precursor of installation art. This interior assemblage was composed of incremental layers of plaster, and discarded and re-purposed found objects in the formation of a fantastical interior with palimpsestic origins.

In contrast, décollage is a French term that literally translates as 'unstick' and the work of Nouveaux Réalistes such as Villeglé and Hains involved '*making art from posters literally ripped from walls*'[10] (Tate, n.d.). Both Hains and Villeglé are renowned for their '*affiches lacerees*', which literally translates as 'torn posters'. Rather than assembling something from scratch in the manner of Schwitters, artists interested in décollage reversed this methodology and instead exploited urban situations rich in billboards, placards and hoardings. Posters would be seized and forensically deconstructed, lacerated and undone to form 'new' historical and political social commentaries. Hains famously referred to himself as an 'inaction painter', inspired by encounters based on the creative 'appropriation' of found objects within a streetscape. Worth noting is that both collage and décollage utilise the palimpsestic devices of layered construction and destruction, or deconstruction and reconstruction (Price, 2016)[11] as an intrinsic aspect of artistic creation.

Another artistic practice that has parallels with the palimpsestic methodologies of collage and décollage is the street art of graffiti. With its guerrilla-like tendencies, appropriation of urban situations and anti-establishment stance, graffiti is in essence a contextually-driven palimpsest. By nature, ephemeral, and available to all as democratised art located in public places operating outside of the perceived elitism of art galleries, graffiti utilises additive tags, caps (larger more elaborate pieces) and stencils. In contrast, the subtractive practice of reverse graffiti creates imagery by removing accumulated dirt and grime. Interestingly, the term graffiti is a corrupted form of the Greek '*graphein*', meaning '*to scratch, draw, or write*'[12] (Art Story, n.d.), and it becomes a chronicler of urban temporality as it literally 'overwrites' or 'unwrites' the experience of the city condition.

Within contemporary art practice Matta-Clark and Whiteread share a preoccupation with the life or biography of a 'found' object. Anyone familiar with the work of Matta-Clark will have already made connections to his artistic oeuvre through a process often referred to as 'anarchitecture', reputedly a conflation of the words anarchy and architecture that, since his death, has become associated with his work. This involved 'unbuilding' abandoned dwellings designated for demolition. His 'Bingo' series involved literal cut pieces taken from a house façade, that, when released from their residential confines revealed the accretions and decorative strata that celebrated the age and imperfections of your typical quotidian domestic environment. Similarly, Whiteread's iconic 'Ghost' involved a concrete cast of the negative spaces of an Edwardian terraced house in East London that was later demolished to reveal these solidified volumes. This inversion of the private realm inevitably immortalised the biography or historiography of the interior. These examples all serve to highlight the life of a building and its associated interior as a form of 'arrested development' that has been recycled as palimpsestic art for our delectation and edification.

The Literary Palimpsest

Figure 2.3 Literary palimpsests typically expose hidden or accumulated layers of meaning through new translations – readings that acknowledge how meaning can alter over time.
Source: Willi Heidelbach, Wikimedia Commons. Licensed under the Creative Commons Attribution-Share Alike 3.0 Unported license: https://creativecommons.org/licenses/by-sa/3.0/deed.en.

The identification of palimpsests within the artistic sphere usefully highlights recurrent themes in relation to the theoretical establishment of commonalities correlating to revisions, revenants, layering, writing and unwriting as both process and evidential traces of change. Literary palimpsests are similarly preoccupied with issues of revelation, but given the context of the written word this typically resurfaces via thematic concerns relating to hidden, accumulated or repressed meanings and interpretations. Dillon's extensive writing on the nature of the palimpsest charts its relevance to literary theory by examining 'textuality' (a process that literally challenges the way we view texts by understanding the perspective of the writer and the reader more fully), a process she refers to as '*risky reading*'[13] (2007: 3). By referencing the writings of nineteenth-century author De Quincey and his use of the term 'involuted', Dillon believes:

> The palimpsest is thus an involuted phenomenon where otherwise unrelated texts are involved and entangled, intricately interwoven, interrupting and inhabiting each other[14].
>
> (Dillon, 2007: 4)

In essence, this process is concerned with an 'accumulation' of meaning, therefore, the process of involution for Dillon is related to textual 'relationality'. This palimpsestic 'contamination' enables readings that reveal deliberately obscure or abstruse interpretations, hidden because of the societal or cultural taboos at the time of writing. Dillon acknowledges her debt to Gilbert and Gubar as their seminal work utilises the palimpsest as a vehicle for exploring the concealed, less socially-palatable meaning hidden within the works of iconic female writers from the nineteenth century:

> ... women from Jane Austen and Mary Shelley to Emily Brontë and Emily Dickinson produced literary works that are in some sense palimpsestic, works whose surface designs conceal or obscure deeper, less accessible (and less socially acceptable) levels of meaning[15].
>
> (Gilbert and Gubar, 1984 [1979]: 73)

By questioning gender equality Gilbert and Gubar not only helped to develop feminist literary theory, but also identified within Victorian literature stories that were awash with archetypal gender stereotypes; those of the mad women in the attic, the witch, the evil queen, even the damsel in distress. These recurring literary tropes were both conformist and non-conformist to the gender, societal and familial constraints of the age (Federico, 2009)[16]. This reinterpretation usefully highlights how the palimpsestic process is utilised within literary theory to reveal hidden narratives relating to history, culture, race, gender and even sexuality (Federico, 2009)[17].

Whilst the impact of the palimpsest as a tool to reveal hidden or repressed narratives is evident in literary theory, the palimpsest can also be utilised to inform 'absence' most notably described in the philosophical work of Derrida's *Of Grammatology*. By taking inspiration from the writings of the philosopher Heidegger (1997 [1974])[18], Derrida developed a discourse concerned with erasure and traces that he referred to as '*sous rature*', or under erasure (Spivak in Derrida, 1997[1974]: xvii)[19]. Within Deconstructivist literary theory, a close reading of a text for Derrida is concerned with '*the undoing, decomposing, and desedimenting of structures*'[20] (1988: 03). It essentially explores meaning by exposing contradictions, acknowledges the debt of time and the importance of translation in 'disrupting' a single, unified interpretation. A text under erasure celebrates the crossed out rather than the permanently deleted. It is not that this crossed out word is incorrect, it has just acquired an 'accumulation' of meaning that the author does not necessarily agree with but cannot readily locate a more suitable replacement. By employing this textual conceit, Derrida believed a deeper legibility for the reader was achieved as the author's intent became transparent. For Derrida these traces recognise revisions or '*différances*'[21] (1988: 85) that lead to an 'expansion' of meaning as the text overtly acknowledges the impact of temporality, variation and change.

This pillaging of literary theory serves to increase our understanding of the palimpsest, its techniques and processes. From Gilbert and Gubar's influential Feminist rereading of iconic Victorian female authors as suppressed voices of rebellion, to Dillon's use of the strategy of risky reading, these examples all illustrate how the palimpsest can be utilised to uncover a repressed, hidden, 'underlying' text. This 'accumulation' of meaning is also highlighted by Derrida's use of the term *sous rature*, as this incorporates a challenge to the original text that is 'editorial' and 'revisionary' in nature. This 'reinterpretation' is possible because palimpsestic temporal layering acknowledges both the past and the present, of the original and the new, in the creation of an 'ongoing' dialogue or narrative. This ability to 'disrupt' the original meaning feels apt given the potential evolutionary nature of an interior palimpsest. This realisation leads naturally to a questioning, an inquiring, as to how these ideas manifest themselves within a more filmic setting.

The Cinematic Palimpsest

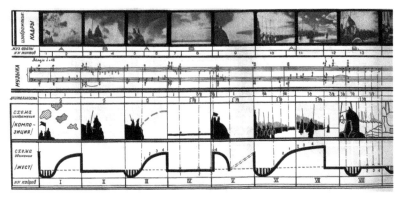

Figure 2.4 Cinematic palimpsests typically chronicle the transitory city condition so the past, the present and the future can coexist on the celluloid screen, embracing montage techniques pioneered by director Sergei Eisenstein. Montage sequence from his film *Alexander Nevsky*.

Source: С.М. Эйзенштейн, Wikimedia Commons. Licensed under the Creative Commons Attribution-Share Alike 3.0 Unported license: https://creativecommons.org/licenses/by-sa/3.0/deed.en.

In examining how the palimpsest manifests itself within a cinematic context, any appraisal rapidly establishes its connection to the milieu of the 'film essayist', unsurprisingly a film genre preoccupied with viewing the city, urban space and humanity as an ongoing narrative. By capturing the ever-changing nature of everyday urban life, cinematically protean streetscapes, framed by mutable, architectural anecdotes and the quotidian lives they contain, become immortalised. It is no great surprise that celluloid palimpsests exploit temporal timelines that 'capture' the constant flux and detritus of urban life within the cityscape. This approach links theoretically to 'Realism' or 'slice of life' cinema, and its celebration and documentation of reality, of the mundane and the everyday. The concept of the 'flâneur' is often associated with this filmmaking oeuvre; a flâneur is defined as an outsider, a passionate but detached observer, someone who operates on the fringes of society who walks the streets observing and recording urban stories. For Jenks this figure becomes '*a multi-layered palimpsest*'[22] (2002: 148) who, by seeing, operates as a reflective mirror that offers an '*alternative vision*'[23] (Jenks, 2002: 149). Synonyms for a flâneur regularly include stroller, dawdler, loafer and loiterer, so a flâneur is often associated with '*the wandering gaze*'[24] (Rascaroli, 2018). This method of dispassionate investigative introspection results in this urban wanderer remaining central to the notion of a cinematic palimpsest 'with filmmakers operating as amateur historians'[25] (Rascaroli, 2018).

This sense of cinematic palimpsests as factually-recorded chronicles is emphatically disrupted by Huyssen's examination of the city condition as he reiterates the importance of the palimpsest to cultural memory, common to both artistic and media practice (2003)[26]. If history is primarily concerned with factual information, Huyssen believes that the vagaries of memory disrupt this. His concept of 'present pasts' is concerned with a shift from the utopian vision of the Modernists who viewed architecture and the city as a positive catalyst for change, appropriately named by Huyssen as 'present futures' to an oppositional view defined by Postmodernism. This stance recalibrated the city condition to one concerned with 'present pasts' whereby both the past and the present are simultaneously in attendance. By re-examining history via historical memory, Huyssen argues that there has been a dissolution of temporal boundaries:

> Historical memory today is not what it used to be. It used to mark the relation of a community or a nation to its past, but the boundary between past and present used to be stronger and more stable than it appears to be today. Untold recent and not so recent pasts impinge upon the present through modern media of reproduction like photography, film, recorded music, and the internet, as well as through the explosion of historical scholarship and an ever more voracious museal culture. The past has become part of the present in ways simply unimaginable in earlier centuries[27].
>
> (Huyssen, 2003: 1)

Cities and the built environment become living, breathing manifestations of temporality, historicity and change; cinematic palimpsests are ideally placed and very adept at capturing this transitory world.

Alternative cinematic interpretations of the palimpsest coalesce in relation to the 'physicality' of a film. By examining the structure of film, questions concerning filmic meaning and perception abound, for example, how films create a shared language of codes and conventions. This realisation encourages the inclusion of filmic 'techniques' to any understanding of the palimpsest, specifically how the final edit is affected via the use of the frame, shot sequences and montage. These techniques all help to 'frame' our understanding of the content, narrative and context of a film as they influence what the audience sees, as well as the sequence in which they see it. The montage as a filmic device is associated with pioneering film theorists such as Eisenstein and Kuleshov, still utilised because it allows for a celluloid spanning of time and space, or a dissolution of present futures as well as present pasts. Its link to a cinematic palimpsest therefore becomes an obvious inclusion, whilst this filmic collation begs the question, what does this analysis of the insidious nature of the palimpsest reveal? From an understanding of the work of the film essayists and their preoccupation with and documentation of cityscapes as ever-changing urban narratives, to the flaneur and their wandering gaze, celluloid palimpsests can be seen to be 'chroniclers' of the city condition, its very history. However, this is not just the case of capturing what is present, but of documenting the past, its very temporality. Huyssen's argument in relation to urban palimpsests sees an intertwining of these two temporal elements, as he believes that '*the past has become part of the present*'[28] (Huyssen, 2003: 1), the former disrupting the latter, and filmic palimpsests certainly share these temporal traits. This relationship to history, memory and the city condition is timely as it starts to occupy the reality of habitable, lived space. There is an obvious overlap with the work of the artists discussed earlier and their fixation on a building's biographical history re-contextualised as art practice, a recognition that begins to 'suggest' how these concerns will manifest themselves within the interior.

The Interior Palimpsest

In exploring the palimpsest's emergence across a variety of creative disciplines, certain traits and characteristics usefully reoccur. This helps to create an investigative structure and a set of expectations in relation to the establishment of a palimpsest that recalibrates its focus to that of the built environment. The term 'interior' palimpsest is deliberate as this epithet is by nature 'inclusive', keeps the focus firmly inside and acknowledges both the value of interior architecture and adaptive reuse alongside more decorative, specifiable expressions of space. It is no great surprise that surface treatment remains integral to a palimpsestic manuscript. Future investigations will demonstrate how this understanding is analogous to the surface treatment of a room, alongside robust architectural modifications and alterations. Scott's influential writings on building adaption echo this stance, believing as he does in two different categories and degrees of alteration from the surface to the spatial (2008)[29]. The

Figure 2.5 Interior palimpsests exploit revisionary and revelatory narratives that consider a building's biography as a perpetual project that looks to both the past and the present. The work of Italian architect Carlo Scarpa captures this temporal alliance within a historic building as a timeline of considered interruptions – Castelvecchio Museum, Verona, Italy.

Source: Alessandro Pace, Wikimedia Commons. Licensed under the Creative Commons Attribution-Share Alike 3.0 Unported license: https://creativecommons.org/licenses/by-sa/4.0/deed.en.

palimpsest's appearance as a theoretical exercise across multiple creative disciplines highlights its significance as a revelatory device; a device concerned with an 'exposé'. Put simply, it becomes a lens that refocuses meaning via interpretation, as the palimpsest theoretically becomes a literal act of 'translation'. This transcription is essentially a 'redrafting' of the original that incorporates alterations and amendments that actively encourages coexistence. The passage of time remains integral to any understanding of a palimpsest as it heralds a burgeoning symbiosis or interdependence in relation to the past and the present, of documentation, retention and alteration as a perpetual ongoing dialogue. These conclusions all contribute towards a greater understanding of how a palimpsest could manifest itself within the built environment.

This investigation now changes its focus and aims to highlight palimpsestic processes that have already divulged their presence (however tentative) within the interior sphere. An initial investigation has already revealed connections to textual analogies in relation to under-, over- and unwriting, characteristics already highlighted as possible shared traits. Machado's oft-cited interpretation of architectural remodelling as another form of palimpsest equally employs textual metaphors to compare the process of remodelling to rewriting:

> . . . remodelling can be seen as writing over, as underlining, as partially erasing, as interstitial writing (writing between the lines), as a way of qualifying, accentuating, quoting, commenting upon, as digression, interlude, or interval, as a way of writing parenthetically, of setting off by punctuation, as a new form for an old story[30].
>
> (1976: 48)

This suggests that the process of alteration or redrafting utilises the strategies of erasing, revealing and rewriting (or under-, over- and unwriting) as an action that becomes potentially crucial to understanding a palimpsestic approach. A response that allows for new involuted risky readings, that disrupt the original and, as Machado states above, and is worth repeating to underline its importance, creates '*a new form for an old story*'[31] (Machado, 1976: 48).

Aksamija, Maines and Wagoner's exploration of archaeologically-important architectural sites are even more explicit with their textual comparisons, as they draw parallels between the creation and revision of a manuscript and a building's construction and subsequent renovation. They embrace the notion of the 'trace' and theorise the formation of the palimpsest via a process of '*temporal layering*'[32] (2017: 9) that is composed of a minimum of four successive chronological episodes, but note that this process becomes indefinite in direct relation to the life cycle of a building. These four key moments become categorised as inscription or creation, erasure, overwriting and recognition (Aksamija, Maines and Wagoner, 2017)[33]. To elucidate further, 'inscription' is concerned with the initial writing or build, whilst 'erasure' as a counterpoint highlights the process of a partial dismantling, removal or hiding of the former. The process of 'overwriting' illustrates the alteration of the original through re-inscription, transforming earlier meaning as a new layer is applied. This, in turn, informs the final stage relating to 'recognition', a process that proffers a new reading as a direct consequence of this alteration. The highlighting of this recurrent textual analogy common to palimpsestic thinking is timely as it employs a transformative process that utilises under- and overwriting, subtraction and addition, to marry, infect, sublimate, celebrate, enhance, but essentially alter, the relationship between the past and the present, the original and the new. This interdependence and architectural alliance gives birth to new ontological meanings, interpretations and translations.

The importance of a building's life cycle and its intrinsic value to an interior palimpsest has been aforementioned, but it is now worth expounding in more detail. If the life of a building is inevitably inscribed on its surface and bound within its spatial and architectural environs, then this condition becomes a 'text'; a text that becomes just another form of architectural storytelling. However, this is not a story of the shiny and new, instead it is a celebration of the battered, the aged and the ramshackle, the historic, the archaeological, even the semi-derelict. This architectural narrative informs the 'biography' of a building, as its history or 'life' is literally writ large on its interior as a series of accretions, incisions, scars and traces. Within this context, Kopytoff's use of the term 'biography' within the field of social anthropology is worth referencing. He extended the notion of a biography, commonly related to people, to 'things' or 'commodities' and this leads naturally to questions regarding life cycle, history, value and worth culturally (1986: 66)[34]. If you can accept that a building has a biography, then this biography inevitably reads as a timeline that encompasses its 'inception' or the point of construction, but it can also acknowledge the many decorative, structural and architectural modifications that occur during its 'life cycle'. This biographical timeline is sustainable in nature, naturally embracing revisions and revenants, employing the oppositional transitory forces of accretion and erosion.

Echoing this burgeoning sense of a temporal biographical approach is Roberts' advocative writings on the 'rehabilitation' of historic and industrial buildings via the auspices of architectural conversion. For Roberts the history of a building, its life cycle, its reuse (usually because of the need to house a new function) is encapsulated by the term 're-utilization'[35] (1989: 4). Re-utilization actively embraces *'conversion as a conceptual instrument in architectural creation'*[36] (Roberts, 1989: 6). An existing building within this context becomes perennially interrupted because it *'can be thought of as an unfinished fragment of a larger edifice'*[37] (Graves cited in Roberts, 1989: 11).

This notion of 'incompleteness' relates to one of the key tenets of the palimpsest by reinforcing the life cycle of a building as a series of endless interruptions. The relationship these interruptions have to the existing condition of the building and its interior becomes a useful avenue for creative revisionary expression and remains central to any understanding of an interior palimpsest and the interrupted interior.

Further evidence for this incipient definition is located in the writings of Brooker and Stone as they also contribute to this debate via their influential publication *Rereadings*. By proposing a theoretical discourse that utilises the terminology of installation, intervention and insertion, they successfully categorise a strategic approach that suggests a palimpsestic 'rereading' of a site in relation to the remodelling of existing buildings (2004)[38]. Site responsiveness when utilising these three strategies becomes an exercise in forging contextual relationships between the new and the old, and of carefully calibrating the level of interdependence between the existing building and any remodelling. Issues of recycling in relation to adaptive reuse naturally recognises a building's biography as '*the memory of the building can be written on its walls, its history pathetically exposed or deeply suppressed*'[39] (Brooker and Stone, 2004: 21). This architectural juxtaposition relishes the oppositional pull of revenants and accretions alongside the ambition to layer a new language and forge a new translation into an existing place.

Stone's evocatively-titled later publication *UnDoing Buildings* explicitly examines the notion of a palimpsest as a device that is concerned with the ongoing development of a building's dialogue, a duality that is preoccupied by both the past and the present. For Stone site-specific narratives or site listening links to an 'activation' of place that remains integral to unlocking any new translation: '*the existing building has a story or narrative that the architect or designer can decipher, interpret and elucidate upon*'[40] (2019: 18). Stone's summation of specific architectural strategies, processes and techniques reinforces a palimpsestic understanding of the built environment as one of constant change embracing as it does memory, addition and subtraction. For the initiated this accumulation of site stories, of layered modifications, enhancements, detritus and disfigurements proffers a rich thread of inquiry. One of the most famous exponents of this approach has to be the Italian architect Carlo Scarpa, as his oeuvre reveals a revelatory preoccupation in relation to a building's inherent narrative. For him the building's history was a codex to be interpreted, prompting a carefully-crafted, surgically-precise response that fused new additions into a layered, stratified whole. This is about the retained, if

Figure 2.6 and 2.7 Zumthor's carefully considered contextual response acknowledges the site's history in its palimpsestic reimagining of a new building built over the remains of a late Gothic church – Kolumba Art Museum, Cologne.

© Raimond Spekking/CC BY-SA 4.0 (via Wikimedia Commons). Licensed under the Creative Commons Attribution-Share Alike 3.0 Unported license: https://creativecommons.org/licenses/by-sa/4.0/deed.en.

deconstructed, layers of a building's skin that is fundamentally different to a *tabula rasa* approach.

Peter Zumthor, winner of the Pritzker Architecture prize, is equally renowned for his sensitively-considered contextual responses. His Kolumba Art Museum in Cologne (1997–2007) is a place-centric response that places a new building over the historic remnants of the late-Gothic Kolumba church, destroyed during World War II. It successfully creates a palimpsestic architectural response that weaves the preserved church ruins and the new contemporary façade of grey brick into a holistic 'tapestried' whole. Similarly, David Chipperfield Architect's

Figure 2.8 The architectural remodelling of the Neues Museum remakes a ruin, developing palimpsestic clues from its lost heritage by evocatively reimagining lost fragments such as the grand staircase – Neues Museum, Berlin by David Chipperfield Architects.
Source: Jean-Pierre Dalbéra, Wikimedia Commons. Licensed under the Creative Commons Attribution-Share Alike 2.0 Unported license: https://creativecommons.org/licenses/by/2.0.

(another Pritzker-winning architect) celebrated remodelling of the Neues Museum in Berlin, in association with Julian Harrap Architects, is a palimpsestic response to a ruined historic building, a direct result of aerial bombardment during World War II. The subsequent visible repairs, conservation and restoration of the original building (1997–2009), after many decades of abandonment, created an opportunity for lost architectural sections to be reinstated, for new infill patches to be installed and for the existing decorative fragments and the obvious discolouration from bomb damage to be integrated into the architectural experience. Complementing this site response was a 'reimagining' rather than a more obvious imitation of the lost grand staircase, an approach that was reinforced by the selection of a new contemporary material. This complex restoration project created '*a new building that, while made of fragments of the old, once again aspires to a completeness*'[41] (David Chipperfield, n.d.). Finally, the architectural practice of Haworth Tompkins (one of their projects will be explored in detail in Chapter Five) is renowned for undertaking an almost archaeological approach to considerations of adaptive reuse in responses that have palimpsestic overtones.

Notions of historical layers, vestige traces and fragments embrace and naturally reinforce the notion of a 'palimpsest'. This process creates a cycle of continual resurrection, reuse and re-inscription, which links to a more decorative analysis of the interior that reveals a preoccupation with surface treatment, traces and authenticity. This relationship to layered historical accretions aligns with the approach of the Rough Luxe Movement and the Japanese philosophies and ideals of Wabi Sabi and Kintsugi. The Rough Luxe Movement analyses a host building, celebrating its patina of age by revealing its historical decorative narrative. Forever associated with exposed strata of wallpaper and paint, with an archaeological fixation that uncovers, retains and celebrates these 'finds', it offsets these deliberately 'undone' interiors with a touch of luxury. Similarly, whilst more than just an aesthetic choice, Wabi Sabi is a philosophy that prizes a simple lifestyle by celebrating the everyday, acknowledges the debt of time via vestigial traces and values '*degradation and attrition*' alongside '*corrosion and contamination*'[42] (Koren, 1994: 28). Both become intriguing phrases relational to the interior condition that heralds an approach that is palimpsestic in origin and celebrates a patina of age and a 'spoilt' or imperfect aesthetic. Kintsugi similarly embraces these concerns by highlighting the scars of fractured objects; remedial work as 'visible mending' becomes celebrated rather than concealed, essentially turned into an art form. These philosophical musings clearly influenced the interiors designed by exponents such as Alex Vervoordt and Rabih Hage, giving rise to a global design doctrine.

Whether you are thinking decoratively or architecturally, the palimpsest as a theoretical exercise can inform multifarious approaches, as taking existing, often-historic building stock, designed for a different purpose that is now defunct, needs careful reimagining. The palimpsestic process highlights the perpetual biography and reuse of a building via an ongoing textual process of underwriting, overwriting and unwriting that deliberately disrupts the original. Retention, revelation, reinterpretation and translation remain central to this burgeoning theoretical stance in relation to interior interruptions. This approach, whilst accommodating architectural change, equally prizes the decorative expression of the interior by valuing the existing qualities of the host as a 'found' recycled aesthetic. When combined with the concerns of adaptive reuse, both have a sustainable agenda that enriches habitation through storytelling.

Understanding what you have, its inherent decorative, architectural, historical and societal qualities, all of which contribute to the building's cultural significance, its very personality, surely informs and enriches any potential interior palimpsest.

Chapter Two Conclusions

In summation, any conclusion aims to establish lessons learnt regarding the characteristics and concerns of a palimpsest by searching for and identifying commonalities. In essence, what defines its nature, recurrent traits or techniques alongside the theoretical and strategic processes it commonly embraces? By deploying an investigative stance across the creative industries, the palimpsests 'tropes' facilitate a series of useful conclusions. A distillation of these conclusions into three key themes results in a realisation as ideas coalesce around the 'original' as a resource, a building's biography and textual techniques. All that remains is to explain this denouement in more detail in relation to the traits that typically embody an interior palimpsest:

1. By viewing the original as a **resource** to be both interrogated and capitalised upon creatively and/or physically, revisions can acknowledge the 'repressed' as well as 'accumulated' layers of meaning.
2. The acknowledgement of a **building's biography** or **narrative**, conceived as an ongoing temporal timeline or perpetual lifecycle of reuse, that rehabilitates the host buildings via an examination of memory and history. This process acknowledges both its past and present incarnations (in fact the two become interdependent) as interruptions in the creation of a future, that exploits its connection and reconnection to place.
3. The use of **textual techniques** or the embracing of textual analogies and terminologies relating to a continual process of temporal 'layering' and 'editing'. This is a system that utilises the oppositional techniques of addition and subtraction, of over-, under- and unwriting, of interruptions as well as disruptions, to 'redraft' the original and rephrase meaning.

A palimpsestic approach suggests an interior no longer needs to be in stasis (although certain historic interiors are notable exceptions) or completely refurbished so that all traces of its previous incarnations are removed. Instead, it can exist somewhere 'in-between'. The palimpsestic process allows a building to occupy the middle ground, neither fully conserved nor completely renovated; instead, it becomes perpetually 'unfinished'. This approach embraces the perennial 'interruptions' as evidenced by a building's biography, a revelatory stance that recognises both additive and subtractive processes, of interruptions and disruptions linked to a <u>fluidity of meaning</u> that alters over time. This understanding leads to important questions such as: how much is interrupted; what is re-inscribed, added to or overwritten; how much is subtracted, erased or unwritten; and what is revealed or reanimated in a building's current translation? This is about making 'value' judgements regarding the biography of a building, decisions that could be guided by conservation or listings considerations, an understanding of its cultural value and heritage, but equally they could acknowledge the possibilities of place-based, site-specific factual site narratives. By consciously making decisions regarding the translation of stories, a building's biography can take centre stage, informing a continual sustainable life. The identification of the qualities that characterise an interior palimpsest via the contradictory auspices of renewal and accretion are now recognised. These two devices aim to give new meaning to adaption through its accumulation of cultural, historical, architectural and decorative detritus and modifications. An interior palimpsest then becomes a revelatory and revisionary experience linked to a building's biography; an experience formed by a series of perpetual interruptions. Forthcoming chapters will explore the 'techniques' regularly employed in relation to these interruptions in more detail.

REFERENCE LIST

1 Britannica. 2022. Palimpsest. *The Britannica Dictionary* [online]. Available at: https://www.britannica.com/dictionary/palimpsest [accessed 24 September 2022].
2 Dictionary.com. n.d. 'Palimpsest'. *Dictionary* [online]. Available at: https://www.dictionary.com/browse/palimpsest [accessed 07 January 2023].
3 Cook, J.W. 2014. 'Palimpsest'. In J.W. Cook, *Encyclopaedia of Ancient Literature* (2nd ed.). Facts On File [online]. Available at: https://search.credoreference.com/content/entry/fofal/palimpsest/0?institutionId=4357 [accessed 7 September 2022].
4 Milazzo, V. 2014. *Encyclopaedia of Ancient Christianity* [online]. Available at: https://search.credoreference.com/content/entry/ivpacaac/palimpsest/0?institutionId=4357 [accessed 7 September 2022].
5 Milazzo, V. 2014. *Encyclopaedia of Ancient Christianity* [online]. Available at: https://search.credoreference.com/content/entry/ivpacaac/palimpsest/0?institutionId=4357 [accessed 7 September 2022].

6 Milazzo, V. 2014. *Encyclopaedia of Ancient Christianity* [online]. Available at: https://search.credoreference.com/content/entry/ivpacaac/palimpsest/0?institutionId=4357 [accessed 7 September 2022].

7 Milazzo, V. 2014. *Encyclopaedia of Ancient Christianity* [online]. Available at: https://search.credoreference.com/content/entry/ivpacaac/palimpsest/0?institutionId=4357 [accessed 7 September 2022].

8 Hollis, E. 2010. 'The House of Life and the Memory Palace: Some thoughts on the historiography of interiors'. *Interiors: Design, Architecture, Culture*, 1 (1–2): 105–117.

9 Tate. n.d. 'Kurt Schwitters'. *Tate Org* [online]. Available at: https://www.tate.org.uk/art/artists/kurt-schwitters-1912 [accessed 10 January 2024].

10 Tate. n.d. 'Décollage'. *Tate.Org* [online]. Available at: https://www.tate.org.uk/art/art-terms/d/decollage [accessed 22 June 2022].

11 Price, N. 2016. 'Contemporary Art Theme: Palimpsests'. *You Tube* [online video]. Available at: https://www.youtube.com/watch?v=9WvBgSx8hW4 [accessed 22 June 2022].

12 The Art Story. n.d. 'Street and Graffiti Art'. *Art Story Org* [online]. Available at: https://www.theartstory.org/movement/street-art/ [accessed 7 January 2023].

13 Dillon, S. 2007. *The Palimpsest: Literature, Criticism, Theory*. Trowbridge, Wiltshire: Cromwell Press Ltd.

14 Dillon, S. 2007. *The Palimpsest: Literature, Criticism, Theory*. Trowbridge, Wiltshire: Cromwell Press Ltd.

15 Gilbert, S.M. and Gubar, S. 1984. *The Madwoman in the Attic: The Woman Writer and the Nineteenth-Century Literary Imagination* (1979). London and New Haven: Yale University Press.

16 Federico, A.R. 2009. 'Introduction'. In Annette R. Federico (ed.). *Gilbert & Gubar's Madwoman in the Attic after Thirty Years*. Columbia and London: University of Missouri Press, 1–26.

17 Federico, A.R. 2009. 'Introduction'. In Annette R. Federico (ed.). *Gilbert & Gubar's Madwoman in the Attic after Thirty Years*. Columbia and London: University of Missouri Press, 1–26.

18 Derrida, J. 1997. *Of Grammatology* (1974). (Translated and preface by Gayatri Chakravorty Spivak). Baltimore and London: The John Hopkins University Press.

19 Derrida, J. 1997. *Of Grammatology* (1974). (Translated and preface by Gayatri Chakravorty Spivak). Baltimore and London: The John Hopkins University Press.

20 Derrida, J. 1988. 'Letter to a Japanese Friend'. In David Wood and Robert Bernasconi (ed.). *Derrida and Différance*. Evanson, IL: Northwestern University Press, 1–5.

21 Derrida, J. 1988. 'The Original Discussion of "Différance" (1968)'. In David Wood and Robert Bernasconi (ed.). *Derrida and Différance*. Evanson, IL: Northwestern University Press, 83–95.

22 Jenks, C. 2002. 'Watching your Step: The History and Practice of the Flâneur'. In Chris Jenks (ed.). *Visual Culture*. Reprinted. London and New York: Routledge, 142–160.

23 Jenks, C. 2002. 'Watching your Step: The History and Practice of the Flâneur'. In Chris Jenks (ed.). *Visual Culture*. Reprinted. London and New York: Routledge, 142–160.

24 Rascaroli, L. 2018. 'The Diachronic Flâneur'. *Mediapolis: A Journal of Cities and Culture* 4 (3) [online]. Available at: https://www.mediapolisjournal.com/2018/10/the-diachronic-flaneur/ [accessed 3 August 2022].

25 Rascaroli, L. 2018. 'Reframing the City of Cinema'. *Mediapolis: A Journal of Cities and Culture* 4 (3) [online]. Available at: https://www.mediapolisjournal.com/2018/10/re-framing-the-city-of-cinema/ [accessed 3 August 2022].

26 Huyssen, A. 2003. *Present Pasts: Urban Palimpsests and the Politics of Memory*. Stanford, California: Stanford University Press.

27 Huyssen, A. 2003. *Present Pasts: Urban Palimpsests and the Politics of Memory*. Stanford, California: Stanford University Press.

28 Huyssen, A. 2003. *Present Pasts: Urban Palimpsests and the Politics of Memory*. Stanford, California: Stanford University Press.

29 Scott, F. 2008. *On Altering Architecture*. London and New York: Routledge.

30 Machado, R. 1976. 'Old Buildings as Palimpsest: Towards a Theory of Remodelling'. *Progressive Architecture*, 57 (11): 46–49.

31 Machado, R. 1976. 'Old Buildings as Palimpsest: Towards a Theory of Remodelling'. *Progressive Architecture*, 57 (11): 46–49.

32 Aksamija, N., Maines, C. and Wagoner, P. 2017. 'Introduction Palimpsests: Buildings, Sites, Time'. In Aksamija, Nadja, Maines, Clark and Wagoner, Phillip (eds.). *Palimpsests; Buildings, Sites, Time*. Turnhout, Belgium: Brepols Publishers n.v., 9–22.

33 Aksamija, N., Maines, C. and Wagoner, P. 2017. 'Introduction Palimpsests: Buildings, Sites, Time'. In Aksamija, Nadja, Maines, Clark and Wagoner, Phillip (eds.). *Palimpsests; Buildings, Sites, Time*. Turnhout, Belgium: Brepols Publishers n.v., 9–22.

34 Kopytoff, I. 1986. 'The Cultural Biography of Things: Commoditization as Process'. In Arjun Appadurai (ed.). *The Social Life of Things: Commodities in Cultural Perspective*. Cambridge: Cambridge University Press, 64–91.

35 Roberts, P. 1989. *Adaptions: New Uses for Old Buildings*. (Translated by Murray Wyllie). New York: Princeton Architectural Press.

36 Roberts, P. 1989. *Adaptions: New Uses for Old Buildings*. (Translated by Murray Wyllie). New York: Princeton Architectural Press.

37 Roberts, P. 1989. *Adaptions: New Uses for Old Buildings*. (Translated by Murray Wyllie). New York: Princeton Architectural Press.

38 Brooker, G. and Stone, S. 2004. *Rereadings: Interior Architecture and the Design Principles of Remodelling Existing Buildings*. London: RIBA Enterprises.

39 Brooker, G. and Stone, S. 2004. *Rereadings: Interior Architecture and the Design Principles of Remodelling Existing Buildings*. London: RIBA Enterprises.

40 Stone, S. 2019. *Undoing Buildings: Adaptive Reuse and Cultural Memory*. New York: Routledge.

41 David Chipperfield. n.d. 'Neues Museum Museumsinsel, Berlin, Germany, 1997–2007'. *Davidchipperfield.com* [online]. Available at: https://davidchipperfield.com/projects/neues-museum [accessed 10 January 2024].

42 Koren, L. 1994. *Wabi Sabi for Artists, Designers, Poets & Philosophers*. Point Reyes, California: Imperfect Publishing.

Palimpsest Techniques, Overwriting   **Three**

Figure 3.1 A palimpsestic interior highlighting 'additive' textual techniques is defined by the co-existence of different temporal eras, expressed as clearly articulated under- and overwriting – Broletto Uno, Mantova, Italy by Archiplan Studio.
Copyright: Davide Galli.

DOI: 10.4324/9781003326267-3

AN INTRODUCTION

Chapter Two highlighted common characteristics, traits and terminologies relating to the palimpsestic process, as well as identifying the pertinence and prevalence of palimpsestic thinking across a variety of creative disciplines. Usefully, this investigative stance culminated in a series of conclusions relational to the establishment of an interior palimpsest. The aim is to now employ and develop this understanding further, for it to evolve, and become further attuned to the preoccupations of the interrupted interior. Integral to this inquisitive journey is for this chapter to expand upon the initial definition of an interior palimpsest, by selecting design precedents that illustrate and deepen the reader's understanding of this type of interior interruption. This chapter consciously moves beyond an appraisal of the characteristics of the palimpsest (already established in the previous chapter) to a categorisation of the 'techniques' it regularly employs. The intention, via an analysis of globally-significant design precedents, is to highlight its pertinence to the built environment, with particular reference to the discipline of interior design, interior architecture and adaptive reuse. Case studies will be analysed against a set of criteria that embeds the conclusions from the previous chapters in order to clarify the assumptions already made. The findings from Chapter One in relation to the importance of narrative via **site-based stories** and responsive place-centric investigations (of an approach linked to **design non-fiction**), alongside the role of **sustainability**, will be duly revisited. The outcomes from Chapter Two in relation to the palimpsests' revelatory and revisionary **narrative** capabilities and its affiliation to a building's **biography** and **textual techniques and analogies** will be utilised to critically analyse selected design precedents in the development of palimpsestic thinking. This is a process that reinforces the original host building as a source of contextual information to be analysed and capitalised upon that becomes a catalyst for change. This recognition helps to establish a forthcoming evaluation centring upon the following criteria, drawn from the outcomes of the previous two chapters:

1. **Project synopsis** – essentially a summary of the project to give much-needed context alongside the rationale for inclusion.
2. **Narrative** – an examination of revelatory and revisionary storytelling via an exploration of ongoing site narratives or design non-fictions in relation to a building's 'biography'.

3. **Textual Techniques** – an exploration of how a palimpsestic approach physically 'redrafts' the existing host building through translational and editorial techniques related to over-, under- and unwriting, interruptions and disruptions.
4. **Sustainability** – an analysis of how sustainable thinking is encouraged by theoretically embracing the concept of an interior palimpsest and its perennial interruptions.

Whilst the importance of both narrative and sustainability has been highlighted in relation to the interior palimpsest, the premise of 'textual techniques' needs expounding. The insidious, close connection of the palimpsest to literary theory is evident as many of the examples included in Chapter Two share this common analogy. Interruptions that ultimately link to a 'redrafting' immediately evoke thoughts in relation to building reuse, as these interruptive practices all draw upon the twin auspices of **alteration** and **retention**, of **addition** and **subtraction**. This duality leads to thoughts on construction, deconstruction, reconstruction and its relationship to a palimpsest in its purest form as an ancient manuscript, as the techniques of erasure, of under-, over- and unwriting echo this process. Whilst there is an obvious difference between the reality of a product such as a book, however historic, and the built environment, certain similarities remain.

This and forthcoming chapters will examine these similarities and their textual allusions and analogies in more detail. The palimpsest as a manuscript is inherently a device that is concerned with surface articulation and close layering, so this investigation will initially explore its application and impact via the expression and coherence of the interior surface. By maintaining this focus, the textual analogy of the interior via the palimpsestic 'devices' or 'techniques' of **writing**, **rewriting**, **overwriting**, **underwriting** and **unwriting** can be explored; especially how they can be utilised to chronicle the 'life' or timeline of any interior. However, interior palimpsests have an obvious advantage over bound volumes and manuscripts in that they exist within a spatial, habitable environment. Within the built environment, the text or narrative can literally 'leap' from the surface, obviously a luxury manuscripts do not share, and one that loosens the close textual relationship to a single page. Instead, physical stratification within the built environment offers both a volumetric and a surface exploration, easing the theoretical bond between the palimpsest's origins and its application spatially. The use of techniques in relation to erasing, highlighting or adding helps to inform ongoing revelatory narratives that harness both design response and site legibility. All of these palimpsestic processes embrace continual reinvention as a series of perennial interruptions via superimposition and involution. The process of writing, rewriting, over- and unwriting becomes an act of sustainable creative translation, of redrafting, concerned with revelation and reinvention.

THE 'TYPE' OF INTERRUPTION – OVERWRITING

If the palimpsest encourages and celebrates interruptions as integral to the act of translation, a key question has to be: what is amplified, what is suppressed and what techniques are employed? By examining how this story is told, commonalities concerning temporal layering, heralded by transitory change, become evident via a revelatory narrative process typically concerned with:

- **ADDITION** – writing/rewriting via overwriting
- **SUBTRACTION** – writing/rewriting via unwriting
- **REDRAFTING** – writing/rewriting via a rewording or rephrasing (or alternatively from interruptions to disruptions)

This chapter, by concentrating upon the palimpsestic '**additive**' technique of **overwriting**, will examine how this technique is utilised to 'reveal', enhance or extend site narratives. Further chapters will explore the 'subtractive' techniques of erasing or removing via unwriting, culminating in translational techniques that 'redraft' and 'reword' the interior narrative. However, a word of caution is necessary, as this 'separation' of techniques utilised to aid comprehension remains simplistic as the boundaries between them remain ambiguous. Indeed, many of the design precedents analysed employ all of these techniques simultaneously. Whilst a more careful reading recognises a hierarchical deployment, analysis and groupings are by nature fluid, concentrating on the primary use of a particular technique to deepen the reader's understanding of the chapter's focus. To facilitate this, five precedents will be analysed within the chapter that examine the interior palimpsest via the techniques it regularly employs. Their selection and arrangement are deliberate, as they progress from an initial surface preoccupation to more robust spatial enhancements or architectural modifications and alterations. In summation, the precedents will advance from the decorative to adaptive reuse, a deliberate collation that informs both the possible scope and the nature of an interior palimpsest. The precedents used are:

- Broletto Uno Apartment, Mantova, Italy – Archiplan Studio
  The renovation of a historic building replete with frescoes creates a palimpsestic residential interior that highlights the co-existence of different temporal eras through the surface application of 'over-' and 'underwriting'.
- Canvas House, Singapore – Ministry of Design Pte ltd
  A traditional shophouse plays with its heritage via a homogeneous painted 'overwriting' punctuated by carefully choreographed 'peek-a-boo' circular reveals, an approach that allows the past and the present to co-exist within this rewritten interior.
- Sala Beckett, Barcelona, Spain – Flores & Prats
  The abandoned old 'Cooperativa pz y Justicia', a former social club, reveals its cultural history and acknowledges its many 'ghosts' via a bricolage of 'under-' and 'overwriting' preserved as collective cultural memory.
- The 10K House, Barcelona, Spain – TAKK
  A project that aims to rethink a traditional residential apartment by questioning sustainable possibilities within strict financial constraints. The introduction of an interior 'house' creates a participatory dialogue of co-existence through occupational 'overwriting' with the host building.
- Astley Castle, Warwickshire, UK – Witherford Watson Mann Architects
  An award-winning scheme that revitalises a historic ruin by 'grafting' a new dwelling directly onto the architectural remnants, this approach to 'overwriting' enhances an ongoing narrative of 'incompleteness' and temporal continuation.

Broletto Uno Apartment, Mantova, Italy – Archiplan Studio

Figure 3.2 and 3.3 Historic frescoes inform a 'found' aesthetic that is both valued and preserved by Archiplan Studio; the original 'underwriting' becomes interrupted by the new decorative 'overwriting', as illustrated by the Broletto Uno Apartment.
Copyright: Davide Galli.

Project Synopsis

Archiplan Studio are clearly a practice immersed in their context; this sense of place ensures they are consistently responsive to the rich history and heritage of Italy. Their studio, located in the Renaissance city of Mantova (or Mantua), Lombardy, is known for and specialises in contemporary interior renovations within historic buildings. Founded in 2000 by architects Diego Cisi and Stefano Gorni Silvestrini, they invariably, given the historical legacy that surrounds them, aim to '*build relationships between the old and the new, creating a balance that achieves a sense of harmony without compromising the innate beauty of the building*'[1] (Cisi cited in Klanten et al. 2019: 83). Intense place-based research informs their many projects, and this investigative approach actively questions contextual responsiveness whilst cultivating a sense of belonging. They refer to their work as '*relational*'[2] (Archiplan, 2021) in that it seeks to explore and build liaisons between the past and the present by balancing the new against the aesthetic canons of earlier eras. Previous interior renovations illustrate this layered approach, such as Apartment G with its vestigial decorative reminders and Apartment RJ with its retained historic decoration.

The inclusion of the Broletto Uno Apartment, completed in 2017, within a chapter devoted to the palimpsestic technique of overwriting encapsulates an approach that embraces decorative evolutionary temporal change via coexistence. Developed as a short stay rental for tourists, the building dates back to the sixteenth century, overlooks the historic Piazza Broletto and encompasses an interior that is notable for its original frescoes. Archiplan's response to these relics resonates with the characteristics and concerns of an interior palimpsest that, whilst occupied with architectural preservation, acknowledges the 'transience of time' and prizes historic stratification. It accommodates unforeseen opportunities between the old and the new, the ancient and the contemporary, in an approach that captures the constant interrupted evolution of an interior's biography.

Narrative

This small provincial city in Northern Italy is an UNESCO World Heritage Site that has become synonymous with the Gonzaga family, a dynasty credited with many of its cultural riches, and this heritage is central to the project's burgeoning narrative. This was a lineage that was also responsible for building many of the city's historic buildings, including the immense and richly-ornamented Palazzo Ducale complex. However, this architectural vibrancy is evident not just in the public buildings, but in the many historic private residences, and the Broletto Uno Apartment is no exception with its richly-frescoed interior. The building's biography reveals a connection to Federico II Gonzaga, who ruled as the Duke of Mantua from 1519 until 1540. Gonzaga was responsible for commissioning the Italian Renaissance painter Romano to paint the frescoes that have survived within the apartment.

Fresco is the Italian word for 'fresh' and frescoes are essentially paintings produced in situ on wet plaster; as the plaster dries the pigment and plaster bond. Employing techniques that have been utilised since the Roman period but are also closely associated to the Italian Renaissance, Archiplan's response to this interior story was one of retention. As they explain: '*We tried to keep together two worlds – the one of the old and the one of the new – in a balance able to guarantee the identity of both of them*'[3] (Archiplan cited in Levy, 2018). Wherever possible the original fresco fragments are preserved and left intact, informing a place-bound revelatory narrative of design non-fiction. This sensitivity to the inherent historical resonance of the building and its interior has resulted in an authenticity that eloquently harnesses the poetics of place in the development of biographic contextual storytelling. This is an approach that emphatically rejects stylistic homogeneity.

Textual Techniques

Figure 3.4 The deployment of palimpsestic textual techniques ensures the retention of the eighteenth-century painted ceiling, complemented by new contemporary fixtures and finishes that enhance biographic contextual storytelling.
Copyright: Davide Galli.

By acknowledging this revelatory narrative, the resultant interior is characterised by its use of textual techniques expressed as historic under- and contemporary overwriting that collectively represent a classic surface dictated palimpsest. Whilst the original fresco remnants are preserved and left intact, the addition of a new layer of plaster or light green paint (a shade chosen because it complemented the colour palette of the fresco) is incorporated. For any frescoes that could not be salvaged, or where the surface was indeed un-frescoed, overwriting was utilised. From a palimpsestic viewpoint the retained patches of frescoes depicting pastoral scenes equates to the original underwriting, whilst the additional new layers of paint and plaster can be interpreted as the new overwriting.

This layered relational approach enhances the experiential quality of the apartment's two principal rooms: the dining room and the main living space. The dining room is characterised by frescoes attributed to the School of Romano depicting a wild boar hunt, complemented by a geometrically-decorated frieze of mythical beasts and an original terracotta floor. An eighteenth-century painted ceiling distinguishes the combined living and sleeping space, created by the demolition of one partition. Included furniture is predominantly bespoke; clearly contemporary, utilising pale timber such as birch or ash, whilst the bathroom and kitchen fixtures and fittings are uncompromisingly modern. Additional pipework is surface-mounted to avoid further damage to the historic fabric of the building. By extending the biographical timeline of the interior, a temporal coexistence is created, whilst a reverence for the past is tempered by the need to create a comfortable vacation home.

Sustainability

The transformation of this heritage interior into a residential apartment for the leisure market is clearly sympathetic to the past, achieved by paying '*particular attention to the historical evolution of the building*'[4] (Cisi cited in Klanten et al. 2019: 83). This evolution is significant as the modified interior timeline ensures previous interior finishes and historic details are respected, repaired and retained. Forming part of a temporal if interrupted continuum that amalgamates the new and the old, historic decorative vestigial traces ensure material consumption is minimised as original, virgin materials are respected. There is, however, a playful resonance with history as other, less important elements are reimagined rather than excised. A bathroom door from the seventies remains as a reminder of past transgressions, whilst the one remaining historic window shutter is reinvented, given new life as part of a wardrobe. These retentions amply illustrate the buildings biographical interruptions that extend beyond its Renaissance inception, a sustainable history that now includes later, contemporary additions and revisions. This 'developmental' strategy ensures the interior is never conceived as static or redundant, but instead is expansive and inclusive, a sustainable stance that Archiplan are clearly attuned to: '*For us, history is a constantly evolving material. We love to explore the beauty of the impermanent, the imperfect, and the incomplete*'[5] (Cisi cited in Klanten et al. 2019: 89).

Canvas House, Singapore – Ministry of Design Pte Ltd

Figure 3.5 By questioning our approach to historic buildings, this recycled interior embraces reuse through homogenous 'overwriting', punctuated by carefully-choreographed circular reveals that offer glimpses into the past – Canvas House, Singapore by Ministry of Design Pte Ltd.
Copyright: Ministry of Design Pte Ltd.

Project Synopsis

Based in Singapore, but with additional offices in Malaysia and China, Ministry of Design (MOD) has garnered numerous plaudits and design awards and is a practice that integrates architecture, interior design and branding into one cohesive vision. Founded by Colin Seah in 2004, the practice philosophy embraces creative inquisitiveness, aiming to always 'question, disturb and redefine'. This inventive approach is evident in an early commission for the New Majestic Hotel that aimed to rethink the hotel experience from (at the time) one of global brand mediocrity to unique consumer experiences linked to locality. Seah explains how the desire to do something different led to this investigative approach:

> The mantra then became question, so ask the deep question about convention, why does convention exist the way it does? Does it need to evolve because time progresses and one needs to find new ways to do old things. To disturb it through design, and then to finally be left with something that redefines . . .[6].

(Annetta, n.d.)

Seah describes his design philosophy as one of 'Essentialism', an approach that distils the elements of *'place, ritual and perception'*[7] (MOD, n.d) and uses these devices to reimagine contemporary situations and experiences. This interest in 'place' when working in global cities that are in a state of constant flux makes for thoughtful, contextually-driven, but playful responses. This irreverent approach, coupled with the desire to rethink typical spatial typologies to transcend convention, goes to the heart of MOD's philosophy.

This questioning agenda is central to the renovation of 'Canvas House' located in Singapore and completed in 2020. This project became an opportunistic reimagining of a historic four-story shophouse exploring how these buildings can 'celebrate' the existing as *'repositories of memories, with previous lives and a past of their own'*[8] (MOD, n.d.). This conscious retention of the building's biography justifies the decision to include this precedent in a chapter devoted to interior palimpsests, as it utilises overwriting to create a blank canvas on which the past and the present become cohabitees. Commissioned by developer Figment, this historic building was repurposed as co-living, long-stay accommodation for ex-pats.

Narrative

Figure 3.6 A place-based temporal narrative informs the approach at Canvas House; conceptually 'time shadows' extend this temporal conceit as the notional shadow cast by the beds allows the past to symbolically interrupt the present.

Copyright: Ministry of Design Pte Ltd.

Canvas House delivers a place-driven narrative (or a work of design nonfiction) that acknowledges the building's biographic history, and this inherent temporality becomes a timeline whereby both past and present collide. By occupying an iconic shophouse (buildings that were originally built for the early immigrant mercantile communities who were central to the city's success as a vibrant trading port) this dialogue could be achieved. Architecturally these buildings typically embraced a Chinese vernacular, with richly ornamental facades, painted in an array of bright colours as live/work spaces. Many of these historic properties were unfortunately demolished until a change of planning policy by the Urban Redevelopment Authority (URA) identified important conservation districts within the city. This organisation provides a useful description of this building type:

> Shophouses – are a prevalent building type in Singapore's architectural and built heritage. These buildings are generally two-to three-storeys high, built in contiguous blocks with common party walls. They are narrow, small terraced houses, with a sheltered 'five foot' pedestrian way at the front[9].
>
> (Urban Redevelopment Authority, 2022)

Canvas House is located in a conservation district known as the 'Blair Plain', a residential neighbourhood near Singapore's Chinatown with the majority of the buildings dating from the 1920's, built in the Refined Rococo Townhouse Style. MOD's response to this rich history was to develop a narrative of homogeneity and continuation.

Textual Techniques

This expressive narrative of continuation was achieved by deploying the palimpsestic textual technique of homogeneous overwriting. MOD covered the internal surfaces, the facade and even the pre-owned furniture in white paint, but left glimpses into the building's biographic past. This carefully-curated 'peek-a-boo' approach to history celebrated original surface treatments and details (the original underwriting) by juxtaposing them against the new, all-white interior (the new overwriting). For Seah, this approach is about the creation of a new blank canvas (hence the name of the project) that aimed to question our reverence for history

Figure 3.7 and 3.8 Existing surfaces, materials, fittings, furniture and even object d'art are reused (if cohesively reimagined via a new white blank canvas) within Canvas House, offset by hints of the original in a curatorial exercise that embraces the palimpsestic textual technique of overwriting.
Copyright: Ministry of Design Pte Ltd.

in favour of embracing a more balanced view, a stance that acknowledges the past but looks firmly to the future:

> When it comes to adaptive reuse projects, the question is always the same, how do we tread the line between the past and the present? . . . If one opts for the project to be just about preservation, it's as good as time standing still, which could be paralysing or inhibiting. But at the same time, neither do we want to disregard history completely by creating something too foreign or novel.[10]
>
> (Seah cited in Levy, 2020)

Instead, this project was about finding a middle ground, neither a hands' off preservation nor a wholesale renovation and contemporary refit. By leaving circular-shaped, unpainted patches as carefully-choreographed vignettes, these scenes act as historic architectural snapshots of the building prior to renovation. These small, scattered, temporal 'interruptions' reveal historic surface treatments such as the brick walls, timber floors and even the materiality of the stair treads, complemented by carefully-selected architectural features such as entrance screens, doors and signs, all offering vivid glimpses into the past. Even the retained, pre-owned furniture and decorative ceramics do not escape this holistic white aesthetic as circular reveals expose the textures, patterns and colours of the original. Conceptually 'time shadows' play with this temporal conceit further as the notional shadow cast by the beds allows the past to symbolically exist in the present. By employing the textual techniques of over- and underwriting, temporal continuity and coexistence is achieved in an approach that is typically associated with an interior palimpsest. These are techniques that are playfully exploited in the creation of a contemporary design response that interrupts by overwriting, that retains, but neutralises the historic past.

Sustainability

This approach gives new life and purpose to a historic shophouse but also acknowledges the limitations on the budget alongside a tight four-month renovation schedule. 'Upcycling' remained central to MOD's rationale as the simple device of a decorative blank canvas (essentially painting the entire house white) is utilised to fast track the scheme to completion ensuring its economic viability. Additionally, these financial constraints led to an upcycling of the existing furniture, interior surfaces and ceramic object d'art. This reuse prompted a sustainable approach of similitude that has helped to both justify and create a neutral interior. MOD's design philosophy accommodates an 'inquiring' approach to heritage and preservation that aims to 'disturb' accepted approaches through temporal layering. In doing so it 'redefines' a response to building reuse that, by embracing decorative homogeneity, becomes wholly sustainable. The notion of layering, longevity and continuation remains central to many palimpsestic schemes and is fundamental to embracing the sustainable mantra of reduce, reuse, recycle, so the creative response undertaken at Canvas House successfully reinforces this ethical agenda.

Sala Beckett International Drama Centre, Barcelona, Spain - Flores & Prats

Figure 3.9 An extensive site investigation celebrates, retains and recycles the decorative remnants of a building's biography, successfully employing the textual techniques of 'over-' and 'underwriting', as illustrated by the rendered section – Sala Beckett, Barcelona, Spain by Flores and Prats.
Copyright: 14. Colour section through vestibule/ Courtesy Flores & Prats.

Project Synopsis
In a chapter concerned with over- and underwriting Ricardo Flores and Eva Prats' response to a former cooperative is a seminal example of architectural remembrance that embraces bricolage, a fortuitous process that creates from whatever is available. Established in 1998, Flores and Prats operate a busy Spanish practice, whilst their experience as both architects and academics has led to numerous publications exploring the discipline of the 'existing'. A co-published volume for IQD serves to underline a shared interest in the 'second hand' and its affinity to architectural reuse, whilst a subsequent publication, Thought by Hand, captures their creative methodology of extensive site documentation via lovingly-produced hand drawings (2022)[11]. Observation and representation as an investigative process is utilised to understand the inherent conditions of a specific site: *'When we sit and draw, we are not looking for a solution, we are trying to understand and recognize the world around a precise project, observing it by drawing . . .'*[12] (Prats, 2021: 24). This contextual sensitivity is central to their renovation of the former Peace and Justice Cooperative building (Cooperativa Paz y Justica) into the Sala Beckett International Drama Centre. The result of a winning competition in 2011, this adaptive reuse project facilitated the rehousing of the 'Beckett Room', a centre that supports creative writing, production and cultivates local talent.

Completed in 2017, 'Sala Beckett', located in Poblenou (a former industrial neighbourhood of Barcelona), prioritises the palimpsestic nature and temporal timeline of a former cooperative, as the building's ghostly biographical remnants remain integral to its architectural resuscitation. The thorough examination, preservation and reinvention of the poetic 'ruination' of the host site is central to its inclusion in this chapter, as over- and underwriting coexists as a form of decorative continuation that acknowledges the importance of heritage within the built environment.

Narrative

Figure 3.10 The building's ruinous narrative via its many 'ghosts' is celebrated, informing the ongoing character of the space in an approach that acknowledges the value of the 'second hand'. Condition of the site prior to renovation.
Copyright: 4. Former café of the worker's club/Courtesy Flores & Prats/Photo by Adrià Goula.

Flores & Prats' obvious veneration for this abandoned building, alongside a celebration of its cultural value and resonance to the community that surrounds it, was integral to their decision not to demolish (even though the building was unlisted). As one of the few remaining examples in the city of a workers cooperative, the building was instead 'rehabilitated'; a process that proved to be sensitive to its accumulated cultural, political and architectural inheritance. By utilising the analogy of a 'used suit' or 'dress', the architects developed an empathetic approach to architectural reuse as one of shared ownership that acknowledges the 'second hand':

> ... our work has something very similar to this form of adaption to a dress already used by others; you have to unstitch it to discover the used pattern, start to work from it, cutting on one side to add on another one, you may need new fabric or to add some pockets. . . . and so on until the garment responds and identifies with the new user[13].
>
> (Flores and Prats, 2019: 1)

This desire to document and understand, to retain rather than discard, led to an investigative analysis of the building's history, sustained by a supportive dialogue with the new client, the director of Sala Beckett Toni Casares. Founded in 1924, its previous incarnation as a worker's community cooperative originally contained a grocery store, café, theatre stage and dance hall until this became economically unviable. Later, unsuccessful uses saw the building reused briefly as both a swimming pool and a gymnasium, followed by 30 years of abandonment. Entering the disused building was akin to stepping back in time, leading to an approach defined by architectural remembrance, that sought to celebrate these collective social memories as 'ghosts'[14] (Flores and Prats, 2020:192). As Prats explains: '*the director told us okay this building is full of stories and Sala Beckett writes new stories so this is encouraging for us to have these around us when we come to work . . .*'[15] (Flores and Prats, 2021).

The ongoing biographic narrative of occupation, abandonment and reoccupation became equitably valued. For three months, Flores & Prats documented the original building, believing explicitly that the process of drawing aided their acquaintance with the site's situational remains. The extensive production of models and drawings aimed to capture the historic, poetic atmosphere, an approach that simultaneously identified architectural fixtures, period details and materials that could be rehabilitated. This extensive site inventory led to a philosophy of reuse and retention that safeguarded vestigial traces, carefully itemised, catalogued and removed during the renovation stage before being reinstated. This is an approach that employs a deliberate narrative of revelatory storytelling, of reclamation and continuity.

Textual Techniques

The notion of both retention and migration is integral to the palimpsestic textual techniques employed to reactivate this interior. This was

Figure 3.11 and 3.12 Flores & Prats embrace palimpsestic textual techniques in a bricolage approach that clearly values the existing as part of a continual evolutionary process at Sala Beckett.
Copyright: 1. First Floor. Lightwell/Courtesy Flores & Prats/ Photo by Adrià Goula.
Copyright: 10. Vesibule at First floor/ Courtesy Flores & Prats/Photo by Adrià Goula.

never about just preserving an abandoned building, of wrapping it in aspic. Instead:

> The ruinous state in which we encountered the building was of interest, not because we wanted to restore it, but rather to take the ruin forward and make it a participant, with its unfinished character of superimposed periods...[16].
>
> (Flores and Prats cited in Murphy, 2017: 116)

This desire to retain and celebrate the past was offset by the inclination to 'add' in the development of an ongoing narrative, surely yet another essential palimpsest trait. The entrance vestibule was enlarged, circulation rethought and the relationship to the original isolated co-operative store on the ground floor addressed. New contemporary additions included the central staircase, lobby seating and a bar as a supplement to those already in existence. A 'patchwork' of historic interior surface treatments (in essence the existing underwriting) became a curatorial exercise depicting a story of perennial evolution. This legibility is made visible via a rationale of decorative overwriting as carefully chosen surfaces are overpainted, but allow historical remnants to remain. This celebratory language of attrition, of investigative preservation, is clearly responsive to the inherent decorative ruination. The 'found' on-site colour palette as a form of overwriting adds visual coherence through red and green wayfinding. This approach created a narrative of past and present that coexisted rather than differentiated (an important distinction) and exploited the notion of an active, continuous life defined by decorative bricolage.

However, this is not just a passive process of editorial retention, it is also an active process of 'cuts', of selective demolition, of erasure and change (and it is worth noting that both processes are integral to a palimpsest, as highlighted in earlier chapters). New oval 'cuts' in the first-floor vestibule create opportunities for light to enter the central circulation spine whilst offering new vistas internally. Inspired by the 'found' ruinous quality of the building prior to its rehabilitation with its open vistas to the sky peppered by light shafts, this remodelling acknowledges the inherited spaces and their lofty dimensions, but sensitively remodels the interior in order to accommodate its new theatrical programme. The embracing of these textual techniques linked to

Figure 3.13 An investigative drawing study analyses the atmosphere and architectural inheritance of the existing building as it searches for clues to inform the colour rationale.

Copyright: 22. Study of colours at entry vestibule/ Courtesy Flores & Prats.

over- and underwriting, of layering, editing, subtraction and addition, exploits palimpsestic techniques and utilises them to articulate the surface as well as the spatiality of the host building.

Sustainability

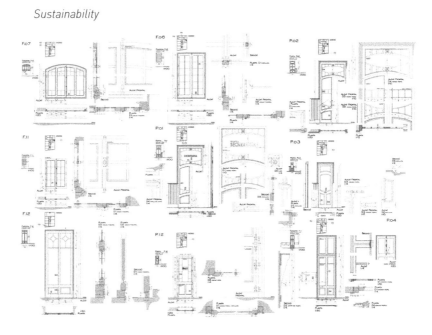

Figure 3.14 A sustainable strategy informs this project via an approach that can be referred to as 'renewed old', as illustrated by the drawing schedule of 'found' doors and windows.
Copyright: 19. Inventory of found doors and windows/ Courtesy Flores & Prats.

The immersive dialogue of succession, of inheritance that resonates within the Sala Beckett, acknowledges its palimpsestic credentials through a conscious decision to celebrate and extend the building's biography. This architectural narrative of continuity, of prolongation, responds to the building's 'past', whilst allowing its 'present' manifestation relating to the changes wrought by its new programme to coexist within the historic building shell. The project response safeguards the building's future by carefully analysing how the existing host building accommodates a new functional programme. Complementing this redemptive architectural approach is the classification of '*renewed old*'[17] (Gómez-Moriana, 2019: 112), which refers to existing elements such as stairs, windows, doors and period flooring that were retained but spliced, mended, resized or migrated to new locations.

This form of architectural rebirth via salvage consciously embraces the circular economy as the building and its interior become locked in a cyclical reuse relationship. By understanding the qualities of the second-hand, by examining the host building, carefully cutting and re-stitching it where necessary, it becomes a good 'fit' for its new community of users. This sustainable realisation is conscious of its many architectural and cultural ghosts, of its apparitional appeal, as the site becomes a treasure trove of finds and discoveries. For Flores & Prats, this biographic response acknowledges the debate regarding how we respond to heritage within a rampant throwaway society by valuing all architectural eras and signs of life equitably. The result is a redemptive architecture, full of residual detritus and traces, that prioritises contextual remains by giving them a new sustainable high '*usage value*'[18] (Lahuerta, 2020: 58). The building's past biography, its preserved fragility, becomes interdependent with the present, inexplicably intertwined in the establishment of a sustainable, palimpsestic response that retains rather than erases, and successfully marries the spirits of the past to the functional needs of the Sala Beckett.

10K House, Barcelona, Spain – TAKK

Figure 3.15 This project embraces evolutionary palimpsestic traits by extending the notion of overwriting beyond its more traditional surface fixation – 10K House, Barcelona, Spain by TAKK.
Copyright: José Hevia.

Project Synopsis

Spanish studio TAKK, founded by Mireia Luzárraga and Alejandro Muiño, have developed an architectural philosophy that resonates with ethical and sustainable concerns which actively embrace design activism. For TAKK, design should always strengthen our relationships to the built environment by 'reformulating' them, a methodology that aims to challenge established practice. Spanning both the public and the residential spheres, their work has won numerous awards, whilst a questioning of thematic concerns relating to gender equality, sustainability and ecology clearly informs their back catalogue. This approach has led to an interrogation of traditional bourgeoisie living arrangements and a reconsideration as to what constitutes taste and decoration. Typical of this approach are projects such as 'The Day after House' an apartment renovation in Madrid that rethinks domestic space by exploring *'new models of use and environmental awareness within the framework of the current energy crisis and climate change'*[19] (TAKK, n.d.).

Their development of the 10K House, located in Barcelona, in 2022 (in reality a 50m² apartment) revisits these concerns in the creation of energy-saving, budget-friendly accommodation that rethinks traditional approaches to renovation. This interrogatory agenda helps to justify this precedent's inclusion in a chapter concerned with the palimpsestic technique of overwriting. Overwriting in this instance transcends the typical surface articulation (as discussed in the previous examples); instead, it becomes a sustained spatial interruption with volumetric credentials. By employing a considered room-size interruption, the characteristics and concerns of an interior palimpsest are evident as the original quality of the host building and its decorative detritus is retained, juxtaposed against its new three-dimensional interior reality.

Narrative

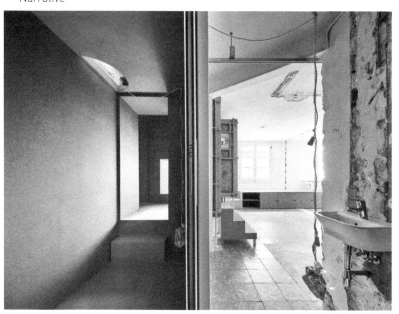

Figure 3.16 Reduced material consumption and financial constraints inform the 10K House's narrative by resonating with the story of the apartment's renovation or deconstruction in an approach that actively challenges traditional residential aesthetics.
Copyright: José Hevia.

The 10K House is sensitive to the existing host building in the development of a narrative of design non-fiction that exploits the duality between old and new, past and present. This biographical fixation ensures that a 'revisionary' temporal narrative is brought into existence, tempered by the desire to renovate for just 10,000 euros, with the clear ambition of being as sustainable as possible. In achieving the project's financial and ethical ambitions, early decisions on site questioned material consumption, informing the incipient interior story. Demolition of the original internal layout resulted in a series of deconstruction traces, with residue remnants delineating the previous location of walls that act as a '*reminder of the original floor plan*'[20] (Englefield, 2023). The twin auspices of economic necessity and sustainability ensured these ghostly revenants were retained, eliminating the need for new materials; instead, these vestigial traces become tangible evidence of previous occupants, of the earlier apartment layout, of its past incarnation. This is a contextual story that is extended by the insertion of a new interior fixture.

Textual Techniques

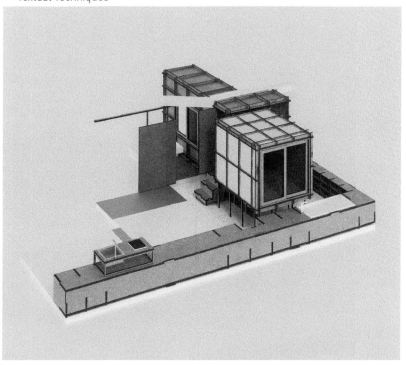

Figure 3.17 Palimpsestic textual techniques within this project became object-based as the newly-inserted fixture (as illustrated by TAKK's model of the 10K House) proffers new surfaces that resonate against the existing host building.
Copyright: TAKK.

The new independent fixture or 'interior house' is a raised structure that is sensitive to 'climatically diverse' interior conditions and functionality. In essence, it becomes an enclosed, warm, cellular space where the rooms 'nestle' into each other within the larger open, if cooler, interior territory of the 'exterior' house. A reductionist approach to materials ensures that the original underwriting (in essence, the host building shell replete with in situ scarification) is retained. This is counterposed against the new overwriting (standard MDF panels and generous amounts of exposed sheep's wool insulation) of the interior house. This textual technique ensures a palimpsestic coexistence of the newly-formed object within and beside the host building that is evident of the apartment's current incarnation. Likened to 'the layers of an onion' or a 'Matryoshka' (a Russian doll that descends in size), this structure, akin to an over-scaled piece of furniture, inhabits the interior, interrupting and modifying its timeline as well as facilitating differing programmatic and temperature requirements. This overwriting at its simplest is just an MDF panel that acts as a sliding door to the bathroom that, when open, becomes a layer juxtaposed against the original wall surface of the host building. However, these panels soon break free of these close confines to form independent three-dimensional spaces. Existing materials remain in situ where possible, integrated into the scheme, just cleaned and polished. By developing this room-based 'object', TAKK ensure that elevational and volumetric palimpsestic opportunities arise as surfaces appear in new spatial combinations, ensuring a continual dialogue of under- and overwriting that progresses from a simple overlain surface to a fully formed three-dimensional reality. Textual techniques in this instance become object-based as the fixture proffers new surfaces that resonate against the existing building in the creation of an interior with palimpsestic overtones.

Sustainability

This project clearly embraces sustainable concerns that acknowledge the climate emergency alongside the current global energy crisis, but which also aim to challenge traditional gendered spaces within the home. The bath is relocated from the darkest, least-ventilated part of the apartment to a space of light on the plan perimeter in the exterior house, whilst the minimal kitchen (similarly positioned) reflects the client's decision to eat out regularly. The adherence to a raw vegan diet negates traditional cooking arrangements, something that can easily be updated if his lifestyle changes. Natural cross ventilation, aided by the removal of the two

Figure 3.18 and 3.19 The new interior fixture or insulated 'house' becomes a warm, cellular space within the cooler interior territory of the 'exterior' house, as highlighted by the thermal gradients and winds plan diagram (Figure 3.18), a stance that is clearly responsive to the global energy crisis. Copyright: TAKK.

internal walls, enhances the liveability of the scheme. The insulated bedroom becomes a space for more than sleeping, as historically these spaces were communal, encouraging a diversity of functions that are sensitive to future living arrangements. The desire to create a home that embraces occupational 'transience', is affordable and is environmentally conscious is obvious. As TAKK explain:

> This causes the heat emitted by us, our pets or our appliances to have to go through more walls to reach the outside. If we place the spaces that need more heat – for example, the room where we sleep – in the centre of the Matryoshka we realise that we need to heat it less because the configuration of the house itself helps to maintain the temperature[21].
>
> (TAKK cited in Englefield, 2023)

TAKK deploy key ethical decisions to achieve this, encompassing thermal efficiency, reduction of material consumption and self-construction.

The design of the new inserted element utilises 'dry' assembly works that eliminate the need for skilled trades and encourages 'non-experts' to engage with the build. The 'house' is fabricated off site, cut using CNC technology and arrives as a flat pack that can be assembled on site using just standard screws. This construction process and the accompanying instruction manual make it eminently flexible, demountable and reusable. The decision to place the fixture on props (in reality recycled table legs), ensures that any services have free passage underneath without the need to chase into and repair existing walls. This also helps to reduce costs, ensuring that the project came in on budget. The integration of such a limited and inexpensive material palette (just MDF panels and local sheep's wool insulation) reduces the carbon footprint of the project. As does an embracing of a raw 'found' unfinished interior aesthetic that harnesses material change and alterations, scars and traces in its reimagining of this apartment's original interior. Evolutionary palimpsestic techniques and concerns are richly evident in the creation of an interior that successfully embeds a sustainable ethos.

Astley Castle, Warwickshire, UK – Witherford Watson Mann Architects

Figure 3.20 The chronological biography of the host building is richly evident as it literally 'grafts' new walls onto a historic ruin, a reoccupation that balances new habitable spaces against a sense of incompleteness – Astley Castle, Warwickshire, UK by Witherford Watson Mann Architects.
Copyright: Philip Vile, courtesy of Witherford Watson Mann Architects.

Project Synopsis

This award-winning project and winner of the Stirling Prize in 2013 (a prestigious award given by RIBA for the UK's best building annually), was concerned with the preservation of an ancient twelfth-century fortified medieval manor house in Nuneaton, Warwickshire that rightly reignited the debate around how, as a society, we consider our architectural heritage. The building's chequered history revealed a series of architectural accretions and uses over time, prior to a catastrophic fire in 1978 that left its future precarious and the abandoned remnants open to the elements for 30 years. The fragility of the ruin necessitated rapid action. An obvious saviour was the Landmark Trust (a building preservation charity), renowned for their work with 'orphaned' architecturally-interesting and historic buildings repurposed as holiday accommodation. The Trust felt that Astley Castle, with its thousand years of constant occupation, was a building of national significance *'associated through ownership with the Grey family with three Queens of England'*[22] (The Landmark Trust: 2012). In 2006, the Trust launched a design competition for invited architectural practices to submit proposals for the rehabilitation of the building. Witherford Watson Mann's (WWM) winning proposal for the rebirth of Astley Castle recognised the inherent conundrum of the potent romanticism of a ruin, tempered by practical considerations of structural stability and the creation of a comfortable, habitable space:

> To place a house inside a ruin, therefore, threatens the essence of each. Two opposite dangers present themselves: the domesticated ruin, which has lost its emotional charge; or the uncomfortable, unsettling house. This was the tightrope we had to walk in making Astley Castle fit for habitation[23].
>
> (Mann, 2016: 7)

Their solution was to consider the building's history as one of active continuance rather than relating exclusively to a passive past. This is a stance that is central to the inclusion of this precedent as it resonates with the concerns of an interior palimpsest as, Janus-like, it looks to the past and the present simultaneously, but views any alterations or change as contributing to an ongoing narrative. The resultant scheme beautifully illustrates how a palimpsest can be spatially responsive beyond the obvious articulation of the surface. By utilising both under- and overwriting, new habitable spaces are accommodated within a ruin that is attuned to its historical resonance. Founded nearly 20 years ago and based in London, WWM are known for projects that aim to comprehend the myriad of built conditions ranging from the cultural to the historic, the urban to the residential. Their work with historic buildings and situations has deservedly brought them global attention and recognition.

Narrative

The project's burgeoning narrative was driven partially by necessity as the cost of renovating the entire ruin was financially prohibitive. The solution was contained in the proposed brief from the Landmark Trust,

Figure 3.21 and 3.22 The project narrative captures the notion of 'half house/half ruin' as only a third of the manor house is re-inhabited, reinforcing the building's evolution in plan as shown by the building's ground floor plan (see Figure 3.21). This approach is echoed in its sectional or three dimensional expression (see Figure 3.22).

Copyright: Figure 3.21: Witherford Watson Mann Architects. Figure 3.22: Philip Vile, courtesy of Witherford Watson Mann Architects.

as this required less usable accommodation; so instead, just a third of the building became enclosed. Strategically this led to the 'secretion' of a smaller house within a larger ruin. For WWM this was an important decision that highlighted both the maintenance of the ruin and its future habitation:

> We have not restored Astley Castle; we have, rather, maintained the ruin and inhabited its core. What is the difference? If restoration implies a form of completion, a return to a past wholeness, we have left the castle incomplete[24].
>
> (Mann, 2016: 8)

This narrative of 'incompleteness' created a powerful biographic relationship between the historic ruin and the new proposed dwelling, the result of extensive site investigations, mapping, drawing and modelling. From this contextual examination, three tactics emerged. Firstly, by understanding the history of the building, its growth or timeline, the decision to locate the new, two-story house in the most historic part of the castle (the twelfth-century core of the original medieval fortified manor house). Secondly, to utilise the fifteenth–seventeenth-century ruins, replete with the domestic remnants of stone fireplaces and mullioned stone windows, as a series of transitory open, partially-roofed courtyards. Finally, the creation of an 'inverted' house with bedrooms on the ground floor and living space on the first capitalised on the far-ranging views through the ruined rooms of the courtyards to the surrounding landscape. By embracing a narrative of ruination and incompleteness, the domestic could become successfully reconciled within the architectural ruin and these two interrelated elements (quintessentially half house, half ruin) could produce a startling symbiosis of temporal loss and gain. This is a synergy that remains central to the notion of an interrupted palimpsestic narrative response.

Textual Techniques

The realisation that history is inexplicably intertwined with our lived condition becoming an active ongoing constituent of the present, rather than separated to some distant past, led to the decision to 'graft' the new house directly onto the ruin. This new graft employs the textual technique of overwriting, illustrating the ongoing continuance of this building's biography. By taking its clues from the previous placement of walls, this architectural 'graft' necessitated '*full contact*'[25] (Mann, 2016: 12),

Figure 3.23–5 Models chart the evolution from the ruined manor house (Figure 3.23), the addition of the stabilising structural armature (Figure 3.24), to the inclusion of internal timber frame of the 'new' house (Figure 3.25). These all serve to capture the interrupted journey of Astley castle and the project's successful use of palimpsestic techniques.
Copyright: Witherford, Watson Mann Architects.

so instead of a house built separately or alongside the ruin, this was a decision based on mutual dependency that had experiential and structural implications:

> Brick clasps the ragged stones, and concrete bridges the great gaps, binding the fragments together, gripping the leaning walls tightly to prevent their fall. Ruin and house are conjoined to the point where each is completely dependent on each other for their existence: old and new hold onto each other for dear life[26].
>
> (Witherford, 2017: 290)

This new grafted layer of brick infill walls and concrete lintels (essentially an armature of new masonry work) becomes a mnemonic revitalisation of the former walls, reforming and rewriting them. This became an exercise in how materials meet and their junctions, with the new bricks carefully accommodating the many irregularities of the ancient masonry walls. This contemporary architectural thread 'stitches' together the remaining ancient fragments, binding the fragile ruin and stabilising it for posterity.

This historical layering gives visual cohesion to the surface treatment of the host building, as rather than repairing years of decay and degradation, gaps and gashes in the building fabric (where possible) are retained. When repairs became essential, brick infill was used, built to the depth, surface variation, difference and scarification of the original walls, while the age and ongoing timeline of the building was celebrated rather than homogenised. This close contextual relationship informs the timber frame of the new house as it nestles into the existing plan geometry with its carefully-crafted rooms of carpentry. These textual techniques all reinforce a narrative of 'incompleteness', creating an aesthetic that acknowledges the incremental interrupted accretion of the 'life' of the castle as a successive process of under- and overwriting.

Sustainability

This exemplary adaptive reuse project successfully rehabilitates a ruined medieval fortified manor house by resuscitating it back to a useful life. Rather than a complete restoration, or a faithful re-creation of what existed previously; the current renovation insidiously intervenes as part of an ongoing 'evolution'. This is an evolution that acknowledges centuries of accretions, accumulations and variations, all representing a cyclical process of growth, decay and rebirth. By embracing continuation, of acknowledging the ongoing biography of a building and accommodating its 'journey', important decisions concerning reuse occur. This approach is inherently sustainable, beginning with an investigatory audit that aimed to understand and evaluate the current site condition. Retention of the existing, whether it be the main structure or its materiality, was the key to unlocking a puzzle that was archaeological in origin. The resultant architecture is akin to geological stratification whereby the processes that have shaped its formation are utilised to guide the design, and are celebrated and displayed for all to see. The approach at Astley Castle is responsive to considered architectural interruptions, and embraces a chronological biography, of a continuous life for the building that actively adds to its history.

Chapter Three Conclusions

This chapter set out to examine the technique of **overwriting** and its central role in the creation of an interior palimpsest, an approach employed to sequentially alter, interrupt and enhance an existing building or space biographically. The biographical fixation becomes central to an evolutionary process that, by recognising the importance of both the original underwriting (or the existing quality decoratively or architecturally) alongside the new overwriting (and its ability to adapt, modify or alter the original), serves to celebrate an additive process that encourages building reuse. Architectural character and idiosyncrasies become valued, prized even, as historic, cultural or societal legacies become celebrated. This analytical exercise serves to highlight the theoretical system alongside its practical application. The five precedents serve to illustrate the range and flexibility of this palimpsestic technique as they progress from surface application, to an enhancement of contextual relationships through to fully-formed, three-dimensional architectural modifications or adjustments. This realisation just necessitates a considered conclusion in relation to what can now be determined from this comparative analysis of to the characteristics and techniques of an interior palimpsest. Does their inclusion reinforce or negate supposed commonalities in relation to the importance of a site narrative or a building's biography, the supposed character and deployment of textual techniques, alongside the embracing of sustainable concerns? This realisation helps to inform and reinforce the following conclusions:

Palimpsestic interiors utilise narrative to:

- Retain and communicate a building's place-centric story by valuing and celebrating the 'biographical' quality of the site and its 'situational' stories.
- Exploit 'revelatory' narratives via decorative and architectural accumulation informed by human habitation as a continuous (if interrupted) occupancy.

Palimpsestic interiors utilise textual techniques to ensure they:

- Integrate the *additive* process of 'layering' or 'overwriting' as part of their evolution.
- Embrace 'transitory' techniques by recognising that any amendments *constantly* revise or 'rewrite' the original narrative.

- Employ 'revisionary' techniques that utilise the additive process of 'layering', of close contact, to introduce overwriting to interrupt the original.

Palimpsestic interiors and sustainability becomes interlinked because they:

- Encourage alteration and retention, and deconstruction and reconstruction, as carefully recalibrated 'relational' dualities that embrace contextual sustainability.
- Value continuance through both revelation and revision, which becomes an expedient sustainable gauge for considering both site responsiveness and reuse.
- Recognise the value of a partial occupation of the existing building footprint as a means to mitigate against operational carbon.

REFERENCE LIST

1 Klanten, A.' Servant, A. and Pearson, T. (ed.). 2019. *The Home Upgrade: New Homes in Remodelled Buildings*. Berlin: Gestalten.
2 Archiplan. 2021. 'Studio'. *Archiplan* [online]. Available at: https://archiplanstudio.com/Studio [accessed 18 September 2023].
3 Levy, N. 2018. 'Archiplan celebrates painted frescoes in subtle revamp of 15th-century Italian home'. *Dezeen* [online]. Available at: https://www.dezeen.com/2018/07/05/brolettouno-apartment-archiplan-painted-frescos-15th-century-apartment/ [accessed 19 December 2022].
4 Klanten, A.' Servant, A. and Pearson, T. (ed.). 2019. *The Home Upgrade: New Homes in Remodelled Buildings*. Berlin: Gestalten.
5 Klanten, A.' Servant, A. and Pearson, T. (ed.). 2019. *The Home Upgrade: New Homes in Remodelled Buildings*. Berlin: Gestalten.
6 Annetta, S. [Presenter]. n.d. 'Design Dialogues: Colin Seah'. *Design Antho/ogy* [online podcast]. Available at: https://design-anthology.com/podcast/2/7/colin-seah [accessed 30 November 2022].
7 MoD. n.d. 'Profile'. *Ministry of Design* [online]. Available at: https://modonline.com/profile [accessed 12 December 2022].
8 MoD. n.d. 'Canvas House'. *Ministry of Design* [online]. Available at: https://modonline.com/projects/canvas-house [accessed 12 December 2022].
9 Urban Redevelopment Authority. 2022. 'The Shophouse'. *Urban Development Authority* [online]. Available at: https://www.ura.gov.sg/Corporate/Get-Involved/Conserve-Built-Heritage/Explore-Our-Built-Heritage/The-Shophouse [accessed 19 December 2022].
10 Levy, N. 2020. 'Canvas House is a co-living space in Singapore with all-white interiors'. *Dezeen* [online]. Available at: https://www.dezeen.com/2020/03/11/canvas-house-singapore-co-living-interiors/ [accessed 19 December 2022].

11 Patlan, S. and Valls, O. (ed.). 2022. 3rd reprint. *Thought by Hand: The Architecture of Flores & Prats*. Amsterdam: Arquine.
12 Prats, E. 2021. 'On Drawing by Hand'. Floresprats [pdf online]. *STOÀ Journal*, 1(2/2): 16–31. Available at: https://floresprats.com/wordpress/wp-content/uploads/2022/03/4.PUB_STUDIO_51.pdf [accessed 20 November 2022].
13 Flores, R. and Prats, E. (guest ed.). 2019. 'Second Hand'. *IQD*, 56: 1 and 58–123.
14 Flores, R. and Prats, E. 2020. *Sala Beckett: International Drama Centre*. Amsterdam: Arquine.
15 Flores & Prats. 2021 'Second Hand: A Virtual Lecture by Eva Prats and Ricardo Flores. [Zoom Webinar by The Design School at ASU]. *You Tube*. Available at: https://www.youtube.com/watch?v=a0h-I7TiZQE [accessed 22 November 2022].
16 Murphy, D. 2017. 'Ghost Storeys'. *The Architectural Review*, 242(1447): 108–17.
17 Gómez-Moriana, R. 2019. 'Circle of Life'. *The Architectural Review*, 1467: 107–15.
18 Lahuerta, J. 2020. 'The New Sala Beckett by Flores and Prats'. In Ricardo Flores and Eva Prats (authors). 2020. *Sala Beckett: International Drama Centre*. Amsterdam: Arquine, 58–63.
19 Takk, n.d. '10K House'. *Takk* [online]. Available at: https://takksarchive.cargo.site/10k-House [accessed 13 September 2023].
20 Englefield, J. 2023. 'Energy-saving 10K House in Barcelona as a labyrinth that multiplies perspectives'. *Dezeen* [online]. Available at: https://www.dezeen.com/2023/03/17/10k-house-barcelona/ [accessed 13 September 2023].
21 Englefield, J. 2023. 'Energy-saving 10K House in Barcelona as a labyrinth that multiplies perspectives'. *Dezeen* [online]. Available at: https://www.dezeen.com/2023/03/17/10k-house-barcelona/ [accessed 13 September 2023].
22 The Landmark Trust. 2012. 'The restoration of Astley Castle'. *YouTube* [online video]. Available at: https://youtu.be/nqjyH08VrwY [accessed 22 December 2022].
23 Mann, W. 2016. 'Inhabiting the Ruin: Work at Astley Castle'. [online pdf], *ASCHB Transactions* 35 [2013]: 5–39. Available at: http://www.wwmarchitects.co.uk/site/assets/files/1225/inhabiting_the_ruin_wwm.pdf [accessed 30 November 2022].
24 Mann, W. 2016. 'Inhabiting the Ruin: Work at Astley Castle'. [online pdf], *ASCHB Transactions* 35 [2013]: 5–39. Available at: http://www.wwmarchitects.co.uk/site/assets/files/1225/inhabiting_the_ruin_wwm.pdf [accessed 30 November 2022].
25 Mann, W. 2016. 'Inhabiting the Ruin: Work at Astley Castle'. [online pdf], *ASCHB Transactions* 35 [2013]: 5–39. Available at: http://www.wwmarchitects.co.uk/site/assets/files/1225/inhabiting_the_ruin_wwm.pdf [accessed 30 November 2022].
26 Witherford, S. 2017. 'Half of it is not necessarily in Ruins'. In José de Paiva (ed.), *The Living Tradition of Architecture*. London and New York: Routledge, 286–96.

Palimpsest Techniques, Unwriting

Four

Figure 4.1 A palimpsestic interior highlighting the subtractive textual technique of 'unwriting' is defined by the partial erasure of the existing space as a form of carefully-considered architectural surgery – HUB Flat, Madrid, Spain by Churtichaga + Quadra-Salcedo.
Copyright: Elena Almagro, courtesy of Churtichaga + Quadra-Salcedo.

DOI: 10.4324/9781003326267-4

AN INTRODUCTION

Chapter Three served to highlight the palimpsestic textual technique of addition as a process that embraced interior interruptions via the inclusion of new, carefully-considered overwriting. This additive process facilitated a grouping of design precedents thematically that investigated and illustrated this approach, whilst acknowledging the commonality of a building's biography when reconsidered as a revelatory narrative that embraces the 'continued' revisionary life of any interior. By exploring this duality of revelation and revision afresh, the techniques of erasure and removal will now be duly examined. This approach acknowledges the transition from the purely additive techniques (explored in the previous chapter) to those that embrace a **'subtractive'** preoccupation. This is a fixation that is still utilised to interrupt an ongoing, perennial site story, charting an evolutionary force that simply **'undoes'** or **'unwrites'**. This is an important distinction, as these revisionary techniques embrace a 'reductive' stance relative to the existing building and its related interior. Temporal incidents expressed as **unwriting** become just another category of interior interruptions integral to the continued, sustainable life of any host building.

THE 'TYPE' OF INTERRUPTION – UNWRITING

If the palimpsest as a theoretical device encourages and celebrates interruptions, this questioning agenda continues to accommodate an investigative interrogation, an analytical exploration of temporal layering and revisionary narratives when deploying the palimpsestic textual technique of:

- **SUBTRACTION** – essentially writing/rewriting via unwriting

This chapter's focus will centre upon how this subtractive process is utilised to reveal, alter or enhance site narratives and embed sustainability into the creative process. To facilitate this, once again five precedents will be analysed in their deployment of this technique. The aim is to build upon the additive techniques of the previous chapter by exploring the oppositional pull of a subtractive agenda. Case study selection is deliberate in its progression from surface excavation expressed as minor cuts, incisions and openings as techniques that directly impact upon the volumetric condition and perceptual experience of space, through to a considered physical 'deconstruction' of the host building concerned with spatial extraction. Consideration of the two textual techniques simultaneously (of overwriting and unwriting) will naturally provide a holistic overview of the palimpsestic process, but as previously mentioned, these have been separated to aid comprehension and highlight key characteristics and traits relative to specific chapters. Central to this investigative journey concerned with unwriting are:

- *The Waterhouse at South Bund, Shanghai, China* — Neri&Hu Design and Research Office
  A series of 'cuts' and 'incisions' into the building fabric 'unwrites' existing spatial relationships, deliberately blurring the boundaries between the public and private realms, conveying the experiential quality of a traditional nong tang or lane house.
- *HUB Flat, Madrid, Spain* — Churtichaga + Quadra-Salcedo
  An examination of 'unwriting' as 'fugues' or carefully punched Matta-Clark-inspired openings into the existing interior reveals new opportunistic spatial relationships.
- *Zeitz MOCAA, Cape Town, South Africa* — Heatherwick Studio
  A seminal adaptive reuse project that embraces the existing character of the building via the 'erosion' of the concrete silos of a former grain store. These carved tubes create new spatial relationships and opportunities via a process of 'unwriting'.
- *Mu.ZEE Art Museum Redesign, Ostend, Belgium* — Rotor
  An interior rehabilitated from the spatial detritus of the previous occupant which employs a reverse methodology that utilises architectural 'surgery' to 'unwrite', sustainably resurrect and repurpose existing interior components.
- *Caritas Psychiatric Centre, Melle, Belgium* — Architecten de Vylder Vinck Taillieu
  An award-winning adaptive reuse project defined by the continued incidental 'deconstruction' of a partially-demolished building whose unexpected stay of execution resulted in its continued life and sustainable evolution.

The Waterhouse at South Bund, Shanghai, China – Neri&Hu Design & Research Office

Figure 4.2 A palimpsestic approach highlighting the subtractive textual technique of partial 'unwriting' utilises 'incisions' to reimagine the experience of the host building exploiting the inversion of the public and the private realms – The Waterhouse at South Bund, Shanghai, China by Neri&Hu.
Copyright: Pedro Pegenaute.

Project Synopsis

Neri&Hu are an interdisciplinary architectural design practice that, whilst responsive to their Shanghai context, operate globally via a series of high-profile projects and an ever-expanding collection of national and international awards. The husband and wife team of Lyndon Neri and Rossana Hu, from the Philippines and Taiwan respectively, founded the architectural office in 2004. The practice encompasses outputs at a variety of scales, typically master planning and architecture, through to considerations of the interior that includes furniture, products, graphics and branding. Their work always questions 'the specificities of program, site, function and history'[1] (Neri&Hu, n.d.) and embraces a fascination with the constant flux and rapid expansion of Chinese cities.

By examining 'place' and place-based heritage, by questioning the loss of local history and culture, the practice has contributed to the debate surrounding the reuse of buildings. This has led to the inclusion of one of their early award-winning schemes, the Waterhouse at South Bund, which was completed in 2010, and is currently operating as a boutique hotel. The existing three-story utilitarian building, built in the 1930s (actually a trio of related properties) and situated on the Shanghai riverfront of the Huangpu, was originally the location of a Japanese Army headquarters. This adaptive reuse project utilises the wider cultural context of the city to question the building's heritage and the traveller's hotel experience as it exploits the palimpsestic traits of erasure, of unwriting, inspired by the communal domesticity of regional vernacular typologies. This strategic approach for Neri&Hu consciously embraces a critical regionalism informed by the writings of Frampton and this place-based sensitivity encompasses the palimpsestic technique of unwriting, justifying its inclusion within a chapter devoted to subtractive erasure.

Narrative

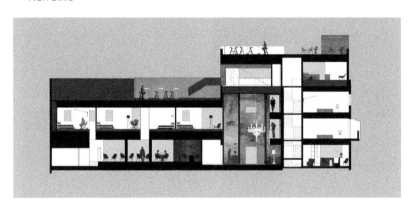

Figure 4.3–4.5 A communal sensibility of shared spaces, adjacencies and site lines aims to give the guest a sense of traditional Shanghainese urban life and informs the hotel narrative as illustrated by the section drawing (Figure 4.3) and the ground and first floor plans (Figure 4.4 and 4.5). Copyright: Neri&Hu.

Neri&Hu's recurring interest in regional vernacular typologies has led them to examine Shanghai's famous nong tang or lane houses, an urban condition that embraced density of living and was typically composed of clusters of buildings connected by a communal lane. The evocative description below gives an insight into the world of this rapidly-disappearing indigenous archetype:

> From the street, these early-20th-century buildings present gabled facades – respectable and a bit staid. But once you walk through the door to the lane running between the houses, you encounter a messy world of clothes hanging out to dry, shutters flung open, people gossiping, and kids running around. Private space bleeds into the public realm, with some folks cooking in the shared lane and others bathing their children there[2].
>
> (Pearson, 2010)

The alternative name for this project is the 'Vertical Lane House', as the communal sensibility just described was integral to the project's burgeoning narrative and this ambition, was less about homage than it was radical reinvention. The writings of Boym and her stance on reflective nostalgia (2001)[3] were another important influence. Nostalgia is a slippery term that can conjure regressive negative connotations, but here it becomes defined by 'belonging' (or even 'longing') via an examination of cultural locality. Instead, Neri&Hu *'embrace nostalgia as a productive lens through which we can consider an alternative reading of historical contexts as well as a different process of design within that context'*[4] (2017: 9) that looks to the past to shape the future. When discussing the project's inspirational inception, the intention was to give the hotel guest a sense of traditional Shanghainese urban life. This was a spatial sensibility that actively questioned the relationship between the public and private realms. This acknowledgement tempered future design decisions relating to the adaption of the existing buildings and the erasure or removal of segments of the host in the creation of this narrative.

Textual Techniques

Unwriting within this project materialises as a series of carefully-orchestrated horizontal and vertical cuts into the existing building's fabric that viscerally translates the story of the nong tang or lane house. Neri&Hu refer to these surgical cuts or interruptions as *'visual slippages'*[5] (Neri&Hu, 2021: 15) and these architectural 'incisions' create unexpected views

Figure 4.6 Neri&Hu develop an approach to openings or 'visual slippages' throughout the building. The courtyard façade utilises corten steel to cover existing apertures, and reclaimed timber panels on the larger windows link to rooms that require privacy, whilst the smaller windows identify spaces that do not.
Copyright: Pedro Pegenaute.

of the private realm within the public sphere by consciously embracing the spatial and visual encounters that the nong tang house traditionally engenders. This is realised via an embracing of the palimpsestic textual techniques of erasure, a device that rather than adding instead subtracts or deletes. In essence, it rewrites by 'unwriting', a process that partially undoes what already exists. By embracing this palimpsestic technique, Neri&Hu's project exploits a *'breaking down (of) the visual, aural and physical limitations of personal space'*[6] (Neri&Hu, 2021: 81) within the hotel that erodes Western concepts of privacy, replacing them with a voyeuristic interplay of *'comfort and discomfort'*[7] (Neri&Hu, 2021: 81).

Initial encounters with these interruptions (or seemingly misplaced windows) occurs in the triple-height entrance lobby, as directly above the reception desk is a deliberately-placed, glazed incision. This vertical cut proffers a view into the supposedly private realm of a guest bedroom, allowing the inhabitant to both see and be seen by the prospective guests in the lobby below. This visual conceit reappears in relation to the hotel restaurant where glazed vertical cuts in the ceiling allow the hotel guests above to be 'obliquely' aware of the bustling dining activity below. The hotel circulation deliberately embraces unexpected vistas both internally and externally, whilst corridors become bridges offering architecturally inquisitorial opportunities. Mirrored window shutters in the courtyard proffer further glimpses into unexpected places, reflecting unforeseen views. The restaurant spills out into a central courtyard, bringing the street directly into the heart of the complex, facilitating further forays into the supposedly private life of the hotel. This technique even permeates some of the guest bathrooms as they sit within glazed boxes exposing the bather to the gaze of others; equally some bedrooms exploit visual connections as glazed opportunities that 'peep' into neighbouring rooms. By carefully undoing through unwriting, the site plays an elaborate, carefully-choreographed game that exploits the interplay and inversion of the public and the private, the communal and the individual, that reinterprets the experience of the lane house. As Neri&Hu explain:

> visual connections of unexpected spaces not only bring an element of surprise, but also force the hotel guests to confront the local Shanghai urban condition where visual corridors and adjacencies in tight nongtang's define the unique flavour of the city[8].
>
> (Neri&Hu, n.d.)

Sustainability

This project, by embracing a profound understanding of place, aims to capture and translate the very essence of Shanghai; this interest in locality ensures Neri&Hu's oeuvre exploits cultural memory. The idea that this contextual understanding reinterprets vernacular typologies via a sensitivity to place is also apparent in their response to the host building. From a sustainable perspective, the building literally becomes a 'found' object (valued even though it was not heritage classified) that is strategically interrupted, refreshed and reused. Decoratively, the retention of surface decay and ruination remains central to the new tectonic reading of the space as '*the real task of this project was to hold back in the restoration process and resist the natural urge to fix every flaw*'[9] (Neri&Hu, 2021: 81). Rather than being covered, patched or ripped out and sent to landfill, the building's biographic history and patina of age was prized, and this inherent site character is archived, retained and recycled. Existing walls remain battered, raw and unfinished, whilst a new structural steel frame reinforces the weakened structure, and new walls are finished in pristine white plaster.

This retention is tempered by the addition of a new fourth level as a roof deck with expansive views across the river to the glittering Pudong skyline, articulated by the use of a new material 'corten steel'. This material (contextually referencing the industrial nature of the working dock) provides tectonic differentiation, its utilisation on the building facade highlighting entrances via canopies and cladding. These new corten layers are literally 'grafted' onto the surface of the host site, whilst an archaeological investigation that is akin to a building autopsy creates opportunities for surgical 'cuts' proffering new spatial, communal and contextual relationships. This approach helped to reactivate the debate regarding the old in a city where the decision to demolish and build afresh are commonplace. This project is about a sustainable retention of the host building tempered by a carefully-controlled deletion; architectural interruptions that embrace subtractive palimpsestic tendencies. However, it is the use of 'incisions' and 'cuts' to create new spatial opportunities and visual slippages that helps to underline the textual technique of subtraction via unwriting, determined by a close reading of, and sensitivity to, the vernacular and cultural traditions of urban life.

HUB Flat, Madrid, Spain – Churtichaga + Quadra-Salcedo

Figure 4.7 A palimpsestic interior highlighting the technique of 'unwriting' via a series of circular cuts or 'fugues' spatially alters but reuses the existing, allowing the past and present to coexist – HUB Flat, Madrid, Spain by Churtichaga + Quadra-Salcedo.

Copyright: Daniel Torrello, courtesy of Churtichaga + Quadra-Salcedo

Project Synopsis

Founded in 1995, Churtichaga + Quadra-Salcedo (ch+qs) are an architectural practice located in Madrid that tackle a diverse range of projects at a variety of scales that consciously rejects specialism. Instead, they exploit creative connections and interdisciplinary overlaps referred to as '*transfusions*'[10] (ch+qs, n.d.), an inquisitive stance that has resulted in a series of national and international awards. The founders, Josemaría de Churtichaga and Cayetana del la Quadra-Salcedo, are interested in the 'intangible' aspects of the built environment that prioritises the experiential and the sensorial. This has led to their involvement in a series of high-profile adaptive reuse projects ranging from a cinema located on the site of a former abattoir, to the reimagining of a pier in Manhattan as an event space. Their practice philosophy is captured by the term '*polytechnic senses*'[11] (Churtichaga, 2014), an attitude that aims to build a relationship between the two apparently disconnected worlds of technique and senses, of the tangible and the intangible. For them, technique relates to the 'physicality' of the built environment, e.g., '*construction methods, structures, science and technical knowledge*', whilst the 'senses' relates to the '*unphysical territory with the way we perceive things*'[12] (Churtichaga, 2014). This preoccupation with intangibles as experiential, perceptual sensations is naturally responsive to the specific atmosphere of a place, resulting in thoughtful, site-sensitive proposals. Their renovation of the HUB Flat, completed in 2011, has been included as an exemplar of this approach (an expansion of the already-successful HUB), in the creation of co-working spaces that aimed to encourage collaborative socially-conscious entrepreneurial projects. The HUB Flat as a precedent resonates with a subtractive palimpsestic process that sustainably reimagines by unwriting or interrupting a pre-existing residential realm.

Narrative

Figure 4.8 and 4.9 A chance find when on site of a photograph of the relative of one of the original occupants (Figure 4.8) informed the project narrative, transforming the space as a place of interconnections via in situ investigations and selective demolition (Figure 4.9).
Copyright: @churtichaga_quadrasalcedo

Narratively the host building's past lives, before a diverse range of former usages left it abandoned and surprisingly untouched (given its location directly opposite the Prado Museum) for many years, was an important source of inspiration. Ranging from the office of a coach station, a garden, to a garage, all indelibly informed the building's biography creating a unique atmosphere, an atmosphere that ch+qs were keen to preserve. This desire led to a project narrative linked to the history of the place:

> We wanted to work more like archaeologists than architects. We have tried not to touch anything because the truly sustainable thing is not to build. If you don't build, you don't consume, you don't spend[13].
>
> (ch+qs cited in Mun-Delsalle, 2011: 96)

This reductionist approach is evident in the renovation of the original Hub (the former garage on the ground floor), as aesthetically and strategically the space remained largely unchanged, except for an upgrade to the mechanical services in relation to heating and cooling. This preservation of the past narratively exploited the building's vintage character alongside an acknowledgement of the inevitable financial and temporal build constraints. The original Hub space could now operate as an event space whilst the adjacent flat, abandoned since the 1950s, was developed to rehouse the co-working spaces. This decision exploited ch+qs's interest in experiential atmospheres and architectural 'disruption' as a means of mediating between the existing and the new; '*disruptive because architecture evolutions and innovations always come from the way we re-read things*'[14] (Churtichaga, 2014). A discovery on site of a photograph of a relative of one of the apartment's previous occupants helped to inform a narrative concerned with geometric cuts and layered interconnections, as the photo with Marco (the builder) holding this person's photograph is both framed and revealed by the new wall opening.

Textual Techniques

By utilising the textual technique of subtraction, inspired by the work of artist Matta-Clark, ch+qs employ this technique as a spatially diagnostic stratagem. The client's desire for a wholly open space, compounded by a lack of budget, inspired a creative response as the antithesis of the existing apartment layout, of cellular rooms arranged around a central spine corridor. ch+qs utilise Matta-Clark's building dissections to reverse the existing condition. By utilising carefully-controlled circular holes or interruptions to create new spatial vistas and opportunities, the footprint and distinctive character of the original apartment was retained. This meticulously-reimagined interior landscape strategically utilised a virtually-modelled cone and two spheres to 'control' the subtractive incidents referred to as 'fugues'. This term refers to its literal Spanish

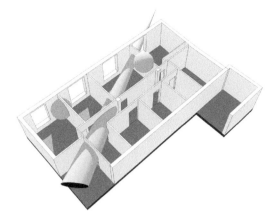

Figure 4.10 By deploying palimpsestic textual techniques, a virtually-modelled cone and two spheres are used to control' subtractive incidents or 'fugues' as illustrated by the strategic diagram.
Copyright: Churtichaga + Quadra-Salcedo.

translation, namely 'extension, an escape or to see beyond', which helps to clarify the placement of these modelled forms as guides to the new openings. This spatial surgery affected everything in its path; mapping openings from the front entrance diagonally across the flat until their exit via the window on the farthest facade. This left two untouched rooms that did not fall within the diagonal line of the cone, and these utilised virtual spheres to map additional circular cuts or fugues.

This approach reinforced the open, collaborative nature expected by the co-working inhabitants as, architecturally, opportunities for aural, visual and spatial dialogues increased as Churtichaga explains:

> Can we keep the memory of the place, the trace of time and make possible a connected space? With a cone from the door to the most distant window and a couple of spheres destroying everything they encountered.... So the space is suddenly interconnected...[15].
>
> (Churtichaga, 2014)

These cuts clearly layer a new palimpsestic language of unwriting via incisions into the host building that highlights a spatial and aesthetic approach which retains but carefully removes, reveals rather than conceals, and plays with depth perception. To reinforce the illusory, perceptual quality, each room decoratively is distinctive, utilising black chalkboard, different painted colour hues or vintage wallpaper from the same era as the apartment to enhance the optical trickery.

Sustainability

Figure 4.11 The palimpsestic approach at HUB Flat embraces sustainability by retaining and then carefully 'fine tuning' the original apartment layout and aesthetic.
Copyright: Elena Almagro courtesy of Churtichaga + Quadra-Salcedo.

By employing the palimpsestic technique of unwriting, ch+qs successfully question stereotypical approaches to building renovation as instead of adding, of embracing the new, they retain. By actively assessing and fine-tuning the existing host building, they exploit a strategy of excavation as interruptions that control the selective demolition undertaken. The act of removal, of cuts and incisions, becomes an aid to the interiors new functional usage and aesthetic sensibility. By carefully undoing or unwriting, rather than redoing, an inherently sustainable approach is realised as proportionately more of the existing building is retained, recycled as part of an active, continuous, cyclical life. The subtractive cuts create new visual and spatial connections essential for its new programme as a co-working space that simultaneously retains the inherent vintage atmosphere of the building, enhancing its unique sense of place. By literally 'mining' the interior and prioritising subtractive excavation, the surface condition is recalibrated, creating extraordinary spatial connections and opportunities that imaginatively rethink a redundant found space. This sustainable transformation actively questions how to renovate without a budget by designing with just 'air' and 'space', and considers how to enhance the volumetric experience of the interior whilst celebrating its vintage character.

Zeitz MOCAA, Cape Town, South Africa – Heatherwick Studio

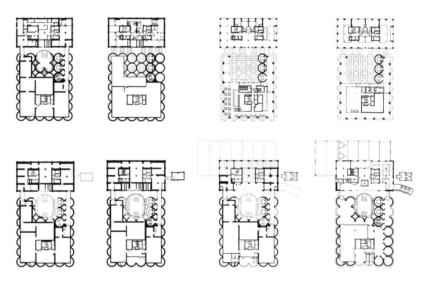

Figure 4.12 and 4.13 The technique of 'unwriting' partially 'erodes' this monumental concrete building, unlocking its rigid geometry via a process of spatial extraction that modifies both the sectional expression (Figure 4.12) and the plan layout (Figure 4.13) – Zeitz MOCAA, Cape Town, South Africa by Heatherwick Studio.
Copyright: Heatherwick Studio.

Project Synopsis

Heatherwick Studio, founded by British designer Thomas Heatherwick in 1994, is renowned for its creative problem solving, interdisciplinary practice and emphasis upon designing by making. A combined studio and workshop in central London ensures that this process is never far from their practice. Heatherwick, a skilled maker himself, is a graduate of three-dimensional design from the Royal College of Art. Working alongside an ever-expanding roster of architects, designers and makers, Heatherwick Studio now problem solves at a range of scales from product and furniture to buildings, infrastructure to urban planning. The studio has won numerous national and international design awards, reflecting its impressive global reach. A profound commitment to sustainability has ensured that the practice always focuses on '*social sustainability, integrating nature and provoking delight, alongside the core issues of materiality, carbon reduction and energy performance*'[16] (Heatherwick Studio, n.d.). A further pledge to 'Architects Declare', essentially architectural practices based in the UK that are committed to resolving the climate crisis and encouraging biodiversity, alongside one to 'Retro First' (a new campaign spearheaded by *Architects Journal*) that aims to prioritise the reuse of existing buildings rather than their demolition and subsequent rebuild, reinforces this stance.

This ethical agenda pervades Heatherwick Studio's oeuvre locally: Broad Marsh based in Nottingham examines the repurposing and reuse of a partially-demolished 1970s shopping mall by prioritising community use and biodiversity. Their award-winning Coal Drops Yard in London regenerates a forgotten industrial hinterland comprised of a series of listed Victorian railway buildings behind Kings Cross station that contributes towards a new retail district and public space. Their distillery and visitor's centre for gin producer Bombay Sapphire in Hampshire was the first refurbishment project awarded an Outstanding BREEAM rating globally. These projects all highlight a commitment to adaptive reuse via the desire to creatively interrogate and reuse existing building stock. This sustainable context helps to explain the inclusion of one of their most iconic adaptive reuse projects, Zeitz MOCAA (Museum of Contemporary Art Africa) located in Cape Town, South Africa. This project, completed in 2011, grappled with the reuse of a former concrete silo and grain store as a new museum for contemporary African art, whilst simultaneously considering how to encourage footfall on a continent with no tradition of museum culture operating in a post-apartheid country. Heatherwick Studio's architectural 'erosion' of the existing concrete silos embraces a palimpsestic process of unwriting via partial erasure that justifies the inclusion of this award-winning precedent within a chapter linked to subtraction.

Narrative

Figure 4.14 The host building (an abandoned grain store and silo) provided a narrative inspired by the story of a single grain of corn, a process that guided the excavation and elliptical sectional expression of the remodelled interior as illustrated by the model.
Copyright: Heatherwick Studio.

Located on Cape Town's historic Victoria & Alfred Waterfront, this project was primarily concerned with the reuse of the heritage-listed Grain Silo complex, an abandoned grain store or grading tower and silo. Built in the 1920s but left abandoned since the 1990s, its position on the city's regenerated harbour earmarked it as an iconic building ripe for development. German entrepreneur Jochen Zeitz, a collector of African Art, alongside curator Mark Coetzee, the former head of the museum, aspired to create a suitable venue for Zeitz's art collection. The appointment of Heatherwick Studio led to the creation of an art museum that aimed to exploit this monolithic concrete structure with a boutique hotel occupying the upper levels. Established through a partnership of the V&A

Waterfront and Jochen Zeitz, the intention was to create a new public not-for-profit cultural institution that focused on collecting, preserving, researching and exhibiting contemporary art from Africa. Whilst keen to retain the building's industrial heritage and ambience, any project solution needed to resolve 'how' to physically connect 42 independent cellular tubes into a cohesive spatial experience. The solution took inspiration from the story of a grain of corn, wholly appropriate given the building's former use as a grain store, an approach that resonated with the building's biography. By digitally scanning this everyday ingredient, the resultant form was rescaled and utilised to 'guide' the erosion and subsequent spatial extraction of the building's formidable internal concrete infrastructure.

Textual Techniques

By employing the palimpsestic technique of unwriting via erasure, the practice's commitment to adaptive reuse is evident. This resulted in a spatial transformation that 'unlocked' the existing restrictive geometry, accommodating the new museum program (opened in 2017, comprising of 80 galleries across nine floors) and gave much needed spatial clarity. The pièce de résistance has to be the atrium, now at the heart of the building, an extraordinary cathedral-like central space that reveals the erosion of eight concrete cylinders three-dimensionally via the elliptical sectional expression of the corn kernel. As Heatherwick explains, this excavation of the existing building was a balancing act '... *a design and construction process that was as much about inventing new forms of surveying, structural support and sculpting, as it was about normal construction techniques*'[17] (2017). Structurally the concrete cylinders were stabilised by casting new concrete sleeves inside the exiting silos; this structural change is evident within the eroded cylinders as aterial lamination. Spatially this erosion helped to accommodate a circulatory strategy of lifts, staircases and footbridges that exploited the newly-created internal views and volume. The resultant ambience acknowledges the palimpsestic quality of the interior, not only via its industrial past (or its original underwriting), but rewrites the present through the technique of unwriting or partial erosion. Together these textual techniques create a temporal marriage that prioritises the successful reuse of the building culturally, programmatically and aesthetically.

Sustainability

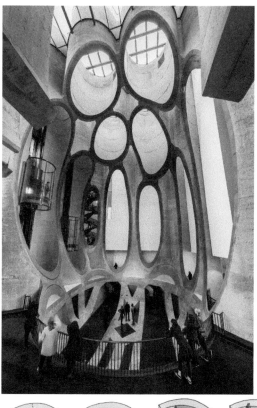

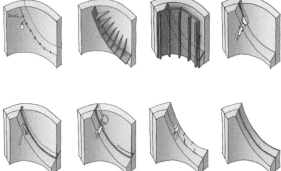

Figure 4.15 and 4.16 The atrium at the heart of this reimagined building reveals the partial erosion of eight concrete cylinders as a creative response to the palimpsestic technique of unwriting (Figure 4.15). Diagram illustrating the process of cutting and structurally strengthening the existing concrete cylinders (Figure 4.16).

Copyright: Figure 4.15: Photo: Iwan Baan. Figure 4.16: Heatherwick Studio.

Heatherwick Studio's sustainable credentials are apparent, as Zeitz MOCAA is clearly a project that prioritises the adaptive reuse of an existing building. The integration of a palimpsestic approach that exploits the technique of unwriting via architectural erosion ensures that the existing character of the building is cherished, linked to a continuous, sustainable life. For Heatherwick, this notion of reuse, of nurturing the ongoing life of a building, exploits an emotional engagement with the things we love, an approach that aims to challenge our prevalent throwaway culture by highlighting affective attachment and the creation of *'out-of-the-ordinary experiences'*[18] (Sudjic, 2020: 5). Within the ongoing climate crisis, the construction industry is currently responsible for 38% of global carbon emissions, whilst the U.K. demolishes 50,000 buildings annually (Heatherwick, 2022)[19]. Heatherwick's response to this problem is a simple sustainable proposition linked to architectural diversity; when you love a building you are less likely to demolish it. This approach is central to Zeitz MOCAA's rehabilitation, an approach that embraces the host site's inherent idiosyncratic character and leads to an innovative strategy. As Heatherwick's explains:

> Just like in nature we have learnt the vast importance of biodiversity, we now desperately need architectural diversity. My goal is to try to help trigger a global humanising movement that no longer tolerates soulless inhuman places. What if our buildings inspired us to want to adapt and adjust and repair them. We can't keep knocking down the buildings around us all the time[20].
>
> (Heatherwick, 2022)

Mu.ZEE Art Museum Redesign, Ostend, Belgium – Rotor

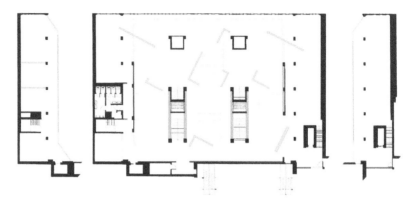

Figure 4.17 By employing a reverse methodology, building obsolescence is questioned; architectural 'surgery' as a process of 'unwriting' repurposes, repositions and reuses the existing interior scenography (highlighted in grey) – Mu.ZEE Art Museum Redesign, Ostend, Belgium by Rotor.
Copyright: Rotor.

Project Synopsis

Rotor are a Belgium-based multi-disciplinary collective renowned for their interest in sustainability within the built environment. Founded in 2006, this ethical ethos is apparent in both their design consultancy and research. Their many influential publications aim to promote material reuse, architectural salvage and reclamation as a means to counter the waste generated by the construction industry. By questioning the ethics of building 'obsolescence', they aim to promote a 'reverse' methodology in relation to construction, renovation and demolition. This ethical 'U-turn' naturally prioritises in situ site audits as inventories that highlight existing materials and architecture details, fixtures and fittings that can be extracted and reused. Their authoritative publication, *Deconstruction and Reuse: How to Circulate Building Elements*, written in conjunction with the Belgian Building Research Institute, examines the circular economy in relation to the building industry and the many challenges it faces.

Rotor co-founder Devlieger writes eloquently on the topic of industrial ecology, is a strong advocate of *'Reverse Architecture'*[21] (2021) and argues for the *'reinstatement of the forgotten art of slowly taking building components apart – when they need to go – in view of their subsequent reuse'*[22] (2019: 37). A sister company, 'Rotor DC' (DC stands for Deconstruction and Consulting), assesses, dismantles, extracts, preps and then resells these carefully-salvaged contextual finds. Hands-on guides such as the 'Reuse Toolkit' (essentially material sheets for the construction industry) highlight common materials that are suitable for 'extraction' or 'reintegration'. Rotor's development of 'Opalis', essentially an online guide that operates as a digital inventory of companies that specialise in salvaging building materials in Brussels, France and the Netherlands, aims to further encourage sustainable specification. Their research has helped to inform both regional policy and EU guidelines in relation to sustainable principles and best practice within the construction industry. This ethical approach is richly evident in the 2021 renovation of 'Mu.ZEE', an exemplary lesson of sustainable reuse that prioritises *'selective dismantling'*[23] (Rotor, n.d.). Rotor, by accessing what is available in situ, undertake a curatorial exercise that carefully unwrites in the development of an ongoing biographical timeline for the host building.

Narrative

Figure 4.18 and 4.19 Rotor's narrative response was concerned with 'recovering' and 'reinstating' the original Modernist quality of the building hidden by numerous internal accretions.
Copyright: Rotor.

Mu.ZEE's narrative addresses the issue of how a former department store becomes reimagined as a museum interior. Originally designed and built as a cooperative by Modernist architect Gaston Eysselinck in 1949, the building embraced many of the tenets associated with this architectural movement. In 1981 the store closed, but the building found new purpose and reopened in 1986 as the Provincial Museum for Modern Art (PMMK). Much later, the art collection from Ostend's Museum for Fine Arts (OMSK) arrived and this became the creative impetus needed to bring this former store back to life. A consolidation of these two collections led to the formation of a new museum entitled Mu.ZEE. This history, typical of many adaptive reuse projects, highlights how a building designed for one purpose ultimately houses another. The task for the project team was to ensure any subsequent renovation was fit for purpose in accommodating its new functional programme. An additional complication, again common to a building that has morphed over time, is the internal accretions and revisions that alter the original layout and litter the interior.

Rotor's ambition was to 'recover' and reinstate the original Modernist quality of the building and this aspiration precipitated an intense period of introversion; in fact, Rotor had just four months to develop a response based upon a close contextual reading of the existing site. An interim period of transition prompted a rethinking of the gallery interiors prior to a major renovation. The resultant project narrative is clear, firstly a teasing out of the story of this Modernist building, its biography and spatiality via a mapping of the modifications or interruptions that have occurred since its inception. This site-bound reading then develops into an exercise in repurposing that embraces a 'second hand' narrative of material reuse.

Textual Techniques

In a chapter concerned with unwriting, Rotor's approach of 'selective dismantling', of a reverse methodology, resonates, as this is an editorial exercise concerned with reductive design. This architectural 'pruning' aimed to reinstate the open, airy, light-filled expanse of the original Modernist interior, characterised by *'a slender skeleton of reinforced concrete (that) supports the three-storey sales area, bathed in abundant daylight that filtered in through huge glass windows'*[24] (Devlieger, 2021: 7). Over time the generous double-height spaces on the lower levels had become concealed, whilst the impressive glass windows that were originally intended were lost. The

Figure 4.20 By employing a process of 'unwriting', of selective dismantling and displacement, the existing rooms are repositioned, rotated and recycled as 'L' shaped partitions, as illustrated by this sketch.
Copyright: Rotor.

imposing concrete staircases and their surrounding open plan spaces had also become enclosed, shuttered and the daylight blocked. Devlieger eloquently captures the degradation of both the ideals and experiential quality of the Modernist interior: '*the building, in which space and daylight originally played a leading role, gradually evolved into a light-shy, fragmented white cube*'[25] (Devlieger, 2021). As an act of design remittance, Rotor, working in conjunction and consultation with the Mu.ZEE team, aimed to 'reverse' this calamitous situation.

A consequence of the reinstatement of the original building quality and a stripping out of the interior was a surplus of building materials and spatial detritus in the guise of former rooms and wall partitions. For a collective concerned with sustainability and circular principles the solution to this conundrum was obvious 'reuse':

> As is often the case with Rotor, the design tries to take what already exists as its departure point: the building, the materials that are released during dismantling; the production capacities of the museum's technical team. We looked for qualities that arise during the removal, rather than the addition of material[26].

(Devlieger, 2021: 7–8)

This ethical approach was integral to the instigation of a new interior 'scenography' (Devlieger, 2021) located on the upper (second-floor) galleries of white freestanding 'walls'. These walls were 'resurrected' from the rooms that previously occupied this floor just *'stripped of their ceilings, then sawn vertically into L-shaped segments'*[27] (Devlieger, 2021: 8). Close consultation with the curatorial team, in-situ testing, as well as the production of scale models, all contributed to precise locations for these revitalised interior elements as a response to site lines between the building and the art. This approach embraces the editorial textual technique of unwriting to repurpose the existing, incorporating it into the new layout and vision for the interior.

Sustainability

Figure 4.21 Rotor develop a sustainable process at Mu.ZEE that considers the interior as a 'found' object ripe for reuse; this addresses the circular economy and material repurposing.
Copyright: Steven Decross.

Rotor's approach to Mu.ZEE highlights how an acknowledgment of a site's history, it's very narrative, can reveal a biographical timeline that celebrates its many modifications or interruptions. This recognition ensures that the process of adaptive reuse is always contextual, as this act of architectural appropriation results in a former cooperative being brought back to active life as a museum. By analysing the continual occupation of a building, a contextual discourse is achieved that redrafts via diagnostic means. Architectural surgery becomes remedial work that exploits change as a positive force, as a therapeutic continuum. By embracing the circular economy and material repurposing, a new interior is literally resurrected from the old. This approach is concerned with a museal environment conceived as a work in progress, a transitory stage that offers the ability to test ideas and approaches prior to a full renovation. Integral to this approach is the recognition, rather than the obliteration, of memory as site incidents and accretions are absorbed and, where possible, repurposed. This curatorial approach results in a gallery that utilises the philosophy and techniques associated with a palimpsest to revitalise and reinvent an interior sustainably. The palimpsestic techniques employed are certainly less obvious than the aesthetic revelatory norm, but as an exercise in revisionary narratives, sustainability and the textual technique of unwriting this creative response offers further insights into the many interruptions common to an interior palimpsest.

Caritas Psychiatric Centre, Melle, Belgium – Architecten de Vylder Vinck Taillieu

Figure 4.22 An unexpected stay of execution for a partially-demolished building resulted in a strategy of controlled 'deconstruction', of further 'unwriting', as a palimpsestic technique to rewrite its future – Caritas Psychiatric Centre, Melle, Belgium by Architecten de Vylder Vinck Taillieu. Copyright: Filip Dujardin.

Project Synopsis

Founded in 2009, Architecten De Vylder Vinck Taillieu (a DVVT) were a Flemish architectural practice composed of Jan De Vylder, Inge Vinck and Jo Tallieu (now disbanded). Their work collectively prioritised 'context' over concept, exploring 'making' as a situational response to the built environment. By encompassing a broad spectrum of differently-scaled outputs from the residential to the commercial, from furniture to building, their thoughtful responses won numerous awards leading to their selection (not once but twice) to represent Belgium at the Venice Biennale. There is a sense in much of their oeuvre of a latent sense of humour, of a playful separation from the ordinary or the everyday as an architectural 'sleight' of hand that creatively exploits the Surrealist heritage of Belgium. Thematically it was their interest in existing buildings via their re-evaluation that has led to the inclusion of their work within a chapter dedicated to the palimpsestic techniques of subtraction. An earlier heritage project titled 'Twiggy' is worth mentioning within this context, memorable for its conversion of a Ghent small city palace (a former single-family house) into a retail store. Operating across a variety of floors, this was a carefully-orchestrated exercise of architectural editing that playfully exploited new surreal spatial opportunities that left a fireplace hanging in mid-air and half a door reimagined as a balustrade. A larger, award-winning project up-scaled these thematic concerns to creatively question approaches to redundant heritage buildings.

The context for this exploration was the Caritas Psychiatric Centre, a residential hospital for patients of all ages with a diverse range of care needs located in Melle, near Ghent. This therapeutic complex, originally built in the 1900s, was composed of a series of separate Flemish Belle Epoque-style buildings or 'pavilions', each designed to house a different department and associated treatment, distributed within an open green park. As these buildings became redundant they were gradually demolished, replaced by more conventional healthcare environments until just two of the original constructions remained. Future plans meant their survival was in doubt but a change in management and the discovery of asbestos on site led to a questioning of the decision to demolish, sadly too late for one of these pavilions but in time to halt the complete demolition of the other. Completed in 2016, this is the story of what happened to this building after this sudden and unexpected stay of execution. By embracing an investigative approach linked to the considered 'deconstruction' of a historic building, this project highlights a holistic strategy of interruptions as controlled subtraction that is central to its inclusion within this chapter.

Narrative

a DVVT were clearly interested in architectural narrative as many of their projects were concerned with how a building's story can be understood, told, developed and retold. For them '*context is so important. Not only context as such, but understood from the widest possible angle . . . Context is an opportunity. An opportunity for change*'[28] (a DVVT, 2019: 27). This intensive site listening linked to a place bound narrative was central to the creative direction developed for Caritas. As one of just three architectural practices invited to contribute to a limited competition, they explored ideas for the retention of the remaining building. The unexpected building stasis (but not prior to it being partially demolished) for a DVVT led to the development of a site narrative concerned with the creation of a new public space or 'square' within the abandoned, partly undone edifice. Their proposal embraced

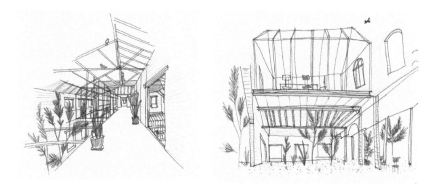

Figures 4.23 and 4.24 Sketches capture an opportunistic approach to building reuse at Caritas Psychiatric Centre as the place-bound narrative exploits the building as an extension of the surrounding park.
Copyright: Architecten de Vylder Vinck Taillieu.

the building as an egalitarian extension of the surrounding public park for the staff, the patients, their families and the public. This resulted in a design response that '... *is like a multiple story park that blurs the boundaries between what is inside and what is outside, what is a building and what is a park, what is a public space*'[29] (Public Space, 2019). This decision is reflective of its sense of place, of retaining the building in its partially-demolished state by recognising the poetic value of its incidental and situational deconstruction. As a DVVT explain, the context informed the burgeoning conceptual response:

> The roof was already gone. The rain reached the ground floor. The wind moved the open windows. But, when the sun came through the old building became a small paradise. Making that small paradise even without the sun. That was the idea[30].
>
> (a DVVT, 2019)

What becomes apparent is that the burgeoning project narrative embraced the beauty inherent in the poetics of building abandonment, of forgotten spaces. It also captures an opportunistic approach to building reuse that incorporates the existing in its partially undone state as the creative impetus for a collective space composed of incidental occupational 'moments' achieved via minimal adjustments.

Textual Techniques

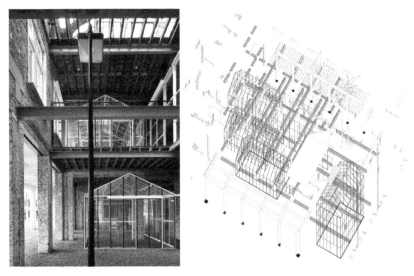

Figures 4.25 and 4.26 By utilising the palimpsestic textual technique of unmaking via 'unwriting', this subtractive approach exploits spatially-ambiguous 'moments' of occupation as illustrated by Figure 4.24. Figure 4.25 illustrates the elements of the design proposal.
Copyright: Figure 4.24: Filip Dujradin. Figure 4.25: Architecten de Vylder Vinck Taillieu.

By utilising the textual technique of subtraction, of further unwriting, this building was successfully brought back to an active life. The removal of the roof covering meant decisions regarding protection from the elements needed urgent resolution. Any interior finishes that could be spoilt by the weather were removed; the ground floor was gravelled, drainage considered, whilst existing building floors were partially demolished to create new vistas within the building volumetrically. Missing windowsills were simply recast in concrete and painted white, whilst the windows on the ground floor were enlarged, becoming permeable thresholds that invited entry. Concrete blocks were utilised for any necessary building repairs to the historic brick envelope, in a deliberate differentiation from the original materiality. The building's 'remains' were stabilised by the addition of a new fluorescent green frame that structurally halted further deterioration. Seven greenhouses were introduced (as notional garden rooms) creating a layer of protection and comfort within this deliberately porous space, their selection reinforcing the conceptual premise of a park within a building that was open to the sky. These decisions all contributed to a loose communal programme of:

> A planted tree. A fireplace. An amphitheatre. Lighting and benches. Jeu de boles and garden chairs. A winter garden and a room for drawing. And all the way to the top, a place to watch the stars[31].
>
> (a DVVT cited in Yoshida, 2017: 81)

This subtractive approach exploits site 'evolution' as the existing building was carefully unmade or unwritten to accommodate its new open condition and loose functionality. This spatial ambiguity accommodated different degrees of enclosure and protection, a necessary precaution given seasonal climatic conditions, whilst proffering a non-prescriptive spatial colonisation composed of gradients of collective space or 'publicness'. Expecting the unexpected thematically is a recurrent theme in a DVVT's work and this project is no exception. Site opportunities, in this instance linked to the building's partial demolition, were celebrated and utilised to unlock the project's creative direction by exploiting contextual clues for reoccupation. This diagnostic architectural unmaking becomes a revisionist palimpsestic textual trait that rewrites by unwriting.

Sustainability

This process of architectural resurrection, of bringing a building threatened with demolition back to a useful life, takes inspiration from its unexpectedly-paused annihilation. This hesitation creates a respite, a hiatus even, that accommodates a rethinking of redundant building stock that embraces architectural reclamation. This approach recognises the intrinsic value of an existing building through a process of subtractive reinvention that accommodates evolution through considered interruptions. An evolution that successfully questions the stereotypical 'conditions' of a building and its future usage. The avoidance of excessive waste as result of wholesale demolition, alongside the evasion of rebuilding costs, helps to minimise carbon emissions, resulting in an economically-viable project, as the building's repair was comparable to the cost of demolition (Murphy, 2018)[32]. Creating spaces that accommodate and protect from different metrological conditions proffers not only an inherent flexibility that boosts the project's occupation, but also negates the associated running costs linked to heating and ventilation. The result is a project that was never about a technical upgrade, instead it advocated a back-to-basics approach that prioritised the physiological and psychological needs of its community of users by questioning how a building can 'change' and evolve. Instead, this 'non-building', with its unexpected transformation of interior spaces into walled outside spaces, is finished without being finished. The value of green spaces and the encouragement of biodiversity as a therapeutic response to mental health is well documented, so the re-imagining of this building as part landscape, part ruin, part interior becomes another aspect of a curative approach that supports and underlines the more conventional treatments on offer. Indeed, the 'remedial' work undertaken throughout the Caritas Psychiatric Centre becomes a metaphor for recovery and recuperation of both its occupants and the building.

Chapter Four Conclusions

In a chapter concerned with examining the palimpsestic technique of **unwriting**, a commonality coalesces around subtraction. What remains open to discussion is how this revelation informs the ongoing examination of the inherent characteristics and techniques of an interior palimpsest. If unwriting rewrites via 'degrees' of erasure, then the included precedents clearly illustrate the progression from smaller incisions or cuts through larger erosions into a considered reimagining of the extant interior scenography, culminating in the considered deconstruction of the existing host building. This introspective realisation cyclically revisits previous questions and assumptions as to whether they reinforce or negate supposed commonalities in relation to the importance of a site narrative or a building's biography, the characterisation and deployment of textual techniques, alongside the embracing of sustainable concerns. This realisation helps to inform and reinforce the following conclusions:

Palimpsestic interiors continue to utilise narrative too:

- Ensure the host building's past lives and its biographic narrative legacy are evidenced as place-bound narratives integral to a new recycled marriage or architectural symbiosis.
- Typically prioritise the retention and/or enhancement of a 'found' aesthetic, as these unique contextual atmospheres have importance creatively and atmospherically to the building's ongoing legacy.

Palimpsestic interiors continue to utilise textual techniques because they:

- Integrate the <u>subtractive</u> process of 'unwriting' into their evolution.
- Embrace an archaeological, almost forensic, unpicking of the existing host building that accommodates evolutionary change.
- Employ 'revisionary' techniques that utilise the subtractive process of 'erasure' to introduce unwriting to interrupt the original.

Palimpsestic interiors and sustainability becomes interlinked because:

- Selective unwriting helps to 'minimise' building waste by considering embodied carbon, as an existing site is instead assessed in terms of its longevity and reuse potential.
- It recognises the building as a 'found' object to be simultaneously retained and pillaged, 'extending' rather than 'ending' the life of a building.

REFERENCE LIST

1 Neri&Hu. n.d. 'About'. *Neri and Hu* [online]. Available at: http://www.neriandhu.com/en/about [accessed 15 February 2023].
2 Pearson, C.A. 2010. 'Project – The Waterhouse at South Bund'. *Architectural Record*. 198(9), [online]. Available at: https://www.architecturalrecord.com/articles/8259-the-waterhouse-at-south-bund [accessed 1 February 2023].
3 Boym, S. 2001. *Future of Nostalgia*. New York: Basic Books.
4 Neri&Hu. 2017. *Neri&Hu Design and Research Office: Works and Projects 2004–2014*. Zurich: Park Books.
5 Neri&Hu. 2021. *Neri&Hu Design and Research Office: Thresholds: Space, Time & Practice*. London: Thames and Hudson.
6 Neri&Hu. 2021. *Neri&Hu Design and Research Office: Thresholds: Space, Time & Practice*. London: Thames and Hudson.
7 Neri&Hu. 2021. *Neri&Hu Design and Research Office: Thresholds: Space, Time & Practice*. London: Thames and Hudson.
8 Neri&Hu. n.d. 'The Vertical Lane House – The Waterhouse at South Bund'. *Neri and Hu* [online]. Available at: http://www.neriandhu.com/en/works/the-vertical-lane-house-the-waterhouse-at-south-bund [accessed 15 February 2023].
9 Neri&Hu. 2021. *Neri&Hu Design and Research Office – Thresholds: Space, Time & Practice*. London: Thames and Hudson.
10 ch+qs. n.d. 'Who we are'. *ch+qs* [online]. Available at: http://www.chqs.net/study.htm [accessed 4 February 2023].
11 Churtichaga, J. de. 2014. 'Polytechnic Senses'. [online lecture]. *YouTube*. University of Toronto, John H. Daniels Faculty of architecture, Landscape and Design. Available at: https://www.youtube.com/watch?v=rnaO73Yys3M. [accessed 4 February 2023].
12 Churtichaga, J. de. 2014. 'Polytechnic Senses'. [online lecture]. *YouTube*. University of Toronto, John H. Daniels Faculty of architecture, Landscape and Design. Available at: https://www.youtube.com/watch?v=rnaO73Yys3M. [accessed 4 February 2023].
13 Mun-Delsalle, Y-Jean. 2011. Hub Madrid. *FuturArc: Asia-Pacific* [online pdf] 23, 94–9. Available at: http://www.chqs.net/archivos/publicaciones/documento_68_0804_110901_futurarc_sep_11_vol_23.pdf [accessed 4 February 2023].
14 Churtichaga, J.de. 2014. 'Polytechnic Senses' [online lecture]. *YouTube*. University of Toronto, John H. Daniels Faculty of architecture, Landscape and Design. Available at: https://www.youtube.com/watch?v=rnaO73Yys3M [accessed 4 February 2023].
15 Churtichaga, J.de. 2014. 'Polytechnic Senses' [online lecture]. *YouTube*. University of Toronto, John H. Daniels Faculty of architecture, Landscape and Design. Available at: https://www.youtube.com/watch?v=rnaO73Yys3M [accessed 4 February 2023].
16 Heatherwick Studio. n.d. 'Sustainability'. *Heatherwick Studio* [online]. Available at: https://www.heatherwick.com/studio/sustainability/ [accessed 3 March 2023].
17 Heatherwick Studio. 2017. 'Unveiled – South Africa's New Contemporary Art Museum', *Press Release* [pdf].
18 Sudjic, D. 2020. 'An Eloquent Path'. In Luis Fernández-Galiano (ed.). *Heatherwick Studio 2000–2020. AV. Monographs*. Madrid: Arquitectura Viva, 222.

19 Heatherwick, T. 2022. 'The rise of boring architecture-and the case for radically human buildings' [online lecture]. *Ted Talks*. Available at: https://www.ted.com/talks/thomas_heatherwick_the_rise_of_boring_architecture_and_the_case_for_radically_human_buildings [accessed 3 March 2023].

20 Heatherwick, T. 2022. 'The rise of boring architecture-and the case for radically human buildings' [online lecture]. *Ted Talks*. Available at: https://www.ted.com/talks/thomas_heatherwick_the_rise_of_boring_architecture_and_the_case_for_radically_human_buildings [accessed 3 March 2023].

21 Devlieger, L. 2021. 'Rotor: Reverse Architecture' [online Lecture]. *YouTube*. The Architectural League and co-presented with The Irwin S. Chanin School of Architecture of The Cooper Union. Available at: https://www.youtube.com/watch?v=6I877eKS8Cg [accessed 22 November 2022].

22 Devlieger, L. 2019. 'Waste Not'. *Architectural Design*, 244(1458), 36–9.

23 Rotor. n.d. 'Mu.ZEE – Art Museum Redesign'. *Rotor* [online] Available at: http://rotordb.org/en/projects/muzee-art-museum-redesign [accessed 18 September 2023].

24 Devlieger, L. 2021. 'Rotor Design Statement'. *Mu.ZEE from Coo to Art* [online pdf]. Available at: https://www.muzee.be/en/getfile/EngelsPersmapVanCoonaarKunstOoiteenwarenhuisnuMuZEE_3045.pdf/bestand [accessed 18 September 2023].

25 Devlieger, L. 2021. 'Rotor Design Statement'. *Mu.ZEE from Coo to Art* [online pdf]. Available at: https://www.muzee.be/en/getfile/EngelsPersmapVanCoonaarKunstOoiteenwarenhuisnuMuZEE_3045.pdf/bestand [accessed 18 September 2023].

26 Devlieger, L. 2021. 'Rotor Design Statement'. *Mu.ZEE from Coo to Art* [online pdf]. Available at: https://www.muzee.be/en/getfile/EngelsPersmapVanCoonaarKunstOoiteenwarenhuisnuMuZEE_3045.pdf/bestand [accessed 18 September 2023].

27 Devlieger, L. 2021. 'Rotor Design Statement'. *Mu.ZEE from Coo to Art* [online pdf]. Available at: https://www.muzee.be/en/getfile/EngelsPersmapVanCoonaarKunstOoiteenwarenhuisnuMuZEE_3045.pdf/bestand [accessed 18 September 2023].

28 Architecten De Vylder Taillieu. 2019. *Variete/Architecture/Desire*. Tokyo: Translated by Kazuko Sakamoto and Yuka Takeuchi. Tokyo: Toto Publishing.

29 Public Space. 2019. 'PC Caritas (Melle, Belguim). Special mention. European Prize for Urban Public Space 2018'. *Public Space* [online video]. Available at: https://www.publicspace.org/works/-/project/k015-pc-caritas [accessed 18 March 2023].

30 Architecten De Vylder Vinck Taillieu. 2019. 'Public Space Panel 3' [online]. *Public Space*. Available at: https://www.publicspace.org/documents/220568/1031395/14523a3_-_2_01.pdf/0714ae45-b566-1213-b9f9-2cd4d40c8adc?version=1.0&t=1525207367398 [accessed 18 March 2023].

31 Yoshida, N.(ed.). 2017. *Architecten de Vylder Vinck Taillieu*. Tokyo: A+U Publishing, 561.

32 Murphy, D. 2018. 'Frame of Mind'. *Architectural Review*. Available at: https://www.architectural-review.com/buildings/frame-of-mind-de-vylder-vinck-taillieus-caritas-psychiatric-centre [accessed 18 March 2023].

Palimpsest Techniques, Redrafting or from Interruptions to Disruptions

Five

Figure 5.1 A palimpsestic approach can embrace both 'interruptions' and 'disruptions' as minor modifications or more major adjustments in its redrafting of an existing building – Battersea Arts Centre, London, UK by Haworth Tompkins.
Copyright: Philip Vile. Courtesy of Haworth Tompkins.

DOI: 10.4324/9781003326267-5

AN INTRODUCTION

As part of this ongoing investigation, further palimpsestic approaches to the built environment will be considered, whilst due emphasis will be given to textual techniques in relation to the process of **'redrafting'**. Whilst previous chapters have highlighted additive techniques in contrast to those that embrace an oppositional subtractive preoccupation, this chapter serves to highlight how new temporal incidents that collectively re-examine and reimagine the interior become just another mechanism integral to the continued, sustainable life of any host building. This form of interior interruption, of architecture amelioration, typically progresses from simple sequential interruptions to a stance that embraces a more decisive holistic disruption; consequently, this chapter consciously oscillates between these responsive gradations. Simply stated, this becomes an exercise in architectural rewriting; in examining **interruptions** and **disruptions** as palimpsestic acts that actively reinforce building reuse.

By taking a palimpsestic approach to a host building, theoretically there is a natural acceptance that contextually-driven site-specific meaning conflates to rewriting via overwriting and unwriting. The previous chapters explored this theoretical stance and utilised precedents to illustrate this approach, collating techniques that can be utilised to highlight, underline or reveal historical and contemporaneous layering. Architectural 'redrafting' becomes just another device that can clarify and extend meaning through new carefully-calibrated temporal incidents or interruptions. Additionally, it can deploy solutions that, through editing, reconsider and substantially 'reorganise' or 'reimagine' the original, a stance that has already been highlighted by some of the precedents in earlier chapters. Redrafting by its very nature is concerned with a 'reappraisal', a rewording of the original that can accommodate a more substantial change and, by extension, an alteration in its meaning or interpretation. This process of alteration, of amending or modifying, **'reworks'** or **'redrafts'** the existing host building via a process that is editorial in origin. The use of the prefix 're' is deliberate as its usage equates to continuous repetition or revision and this evolutionary stance becomes both a founding principle as well as a recurring trait.

THE 'TYPE' OF INTERRUPTION – REDRAFTING, FROM INTERRUPTIONS TO DISRUPTIONS

If the palimpsest as a theoretical device encourages and celebrates interruptions as integral to the act of translational alteration, further interrogation is beneficial in relation to the ongoing question of the what, how and why of this biographical narrative. Focus will now centre upon how transitory change is architecturally expressed, especially separation, clarification and difference within the built envelope. This book's central proposition remains consistent, allowing for a continued investigation of temporal layering, revelatory, cumulative and revisionary narratives via the deployment of the palimpsestic techniques. This acknowledgement accommodates an examination and chapter-focus centred upon:

- **REDRAFTING** – writing/rewriting via rephrasing, or alternatively from interruptions to disruptions

By focusing on the palimpsestic techniques of rewriting via redrafting, an examination of how existing site narratives are enhanced, reimagined, altered or adjusted will follow. To facilitate this once again, five design precedents will be analysed against a set of previously-established criteria, interrogated for further insights into the formation of an interior palimpsest. The aim is to build upon the additive and subtractive techniques of the previous chapters by highlighting additional palimpsestic possibilities and opportunities. Precedent selection is deliberate, choosing examples of redrafting that illustrate a sequential approach that begins by illustrating more minor interruptions or 'modifications' such as repairs or patches as partial alterations to disruptions that redraft, or rewrite, through more significant 'adjustments' to the building fabric, characterised by a forcible separation or division. Both palimpsestic techniques help to 'sequence' or even 'resequence' a story, and clarify meaning by highlighting and expressing the passage of time.

These techniques can just relate to the tectonic expression of the interior, but equally they can inform or alter the existing building shell, challenge the spatial and volumetric organisation, aim to enhance the building's usage, as well as inform the overall perceptual expression of the host building. This form of architectural and interior-centric evolutionary adaption takes a curatorial approach to the physical site that typically repairs, reimagines and reworks existing content. Consequently, the precedents included illustrate redrafting in a successive, progressive manner as they proceed from interruptions to disruptions, from minor modifications to major adjustments. They are:

- *Selo Shoe Shop, São Paulo, Brazil* – MNMA
  This interior exploits material degradation as a kintsugi-inspired incident, the 'visible' repair 'interrupts' the site narrative modifying it afresh.
- *Maison de l'Architecture, Paris, France* – Chartier-Corbasson Architects
  New corten steel 'patches' make legible a series of pauses that punctuate and 'interrupt' the chronological historical strata that characterises this building adaption.
- *Battersea Arts Centre, London, UK* – Haworth Tompkins
  A listed civic building, formerly a town hall, inspires a series of site-specific evolutionary 'adjustments' that enhance the interior occupation as well as 'disrupt' and reimagine the lost roof of the Grand Hall after a devastating fire.
- *Sedgwick Rd., Seattle, Washington* – Olson Kundig Architects
  This 'disruption' to a heritage building utilises a 'Frankenstein' revisionistic approach in its 'adjustment' of historic fragments, of something new being created from the sum of the original historic parts.
- *Military History Museum, Dresden* – Studio Libeskind
  The adaption of this historic building deliberately 'disrupts' both its classical symmetry and its history via an architectural schism that redrafts and 'adjusts' the experiental quality of the site.

Selo Shoe Shop, São Paulo, Brazil – MNMA

Figure 5.2 This palimpsestic interior embraces a site flaw as a Kintsugi-inspired 'visible' repair or modification that 'interrupts', but ultimately embellishes, the existing site narrative – Selo Shoe Shop, São Paulo, Brazil by MNMA.
Copyright: Andrè Koltz.

Project Synopsis

As a response to a problematic site, this project illustrates the power of 'repair' as integral to the ongoing story of an interior, a narrative developed by Brazilian-based architectural practice MNMA (pronounced 'minima'). Founded by Marian Schmidt and André Pepato in 2014, the practice's name captures their architectural sensibility, developed by simply deleting unnecessary letters from the word 'mínima' with the intention of '*mak(ing) the smallest possible configuration that would still convey the idea*'[1] (Ingram, 2021: 26). Based in the frenetic city of São Paulo, their architectural oeuvre has developed via a series of small-scale retail projects that are remarkable for their pared back, reductive sensual sensibility. A careful reading of the inherent site condition has resulted in a contextual problem becoming an active constituent of the project solution, whilst their celebration of and desire to engage with vernacular materials and artisan traditions allows for the exploitation of everyday mundane materials in an innovative and celebratory manner. This approach has been tested and developed via a series of high-end boutique interiors for Roupateca, Egrey, Haight and Selo.

The Selo Store in São Paulo, completed in 2019 and winner of the Prix Versailles Special Prize, is of especial interest within the context of this chapter, as it employs an approach that draws parallels with and reciprocates the concerns of an interior palimpsest. By utilising the Japanese technique of 'kintsugi', the project proposal celebrates inherent site problems by making them integral to a deeper understanding of the interior's biography. This approach justifies this precedent's inclusion in a chapter devoted to temporal site interruptions that aims to 'redraft' site meaning via a simple incident that punctuates and 'modifies' the interior.

Narrative

Figure 5.3 and 5.4 MNMA have delivered a project narrative linked to locality, not only as a space of introversion and silence as an antidote to the city condition, but also as a spoilt material narrative concerned with temporality.
Copyright: Andrè Koltz.

The Selo store, a boutique selling handmade shoes, captures MNMA's minimalistic approach of a material-led project solution that is responsive to its locality. The desire to create a pause, a moment of silence as a response to the city condition, is a recurrent theme in their work as Schmidt explains:

> In my opinion we create spaces where you can be alone and experience solitude. I like the word solitude because it's not the same thing as loneliness. Solitude is an incredible feeling whereas loneliness is negative . . . But they are places in which you can simply breathe[2].
> (Schmidt cited in Ingram, 2021)

This quasi-spiritual approach of simple interior volumes and forms, a stripped back materiality and a judicious use of light results in elemental, restful interiors. This experiential narrative is evident within the Selo Store as the façade and entry as a negative volume seamlessly echoes the materiality of the street, utilising white cement slates and frameless glass in the formation of this transitional space. Internally a circular skylight inspired by the artist James Turrell (famous for his perceptual experiments with light and space) complements the pervading atmosphere, creating a connection to the sky and illuminating the naturally-pigmented textural concrete wall render replete with build striations. Bespoke cast furniture utilises the same material complemented by a cylindrical volume that creates a contemplative space of simplicity, intimacy and silence as a space of concentric introversion. However, the floor narrative resonates with palimpsestic concerns, as the story of site repair becomes an accidental incident or interruption within the interior. The inherited host space is inclusive of cracking on the existing concrete floor, a consequence of material expansion over time that needed to be addressed. Rather than replace or conceal this evident site rupture, this material error is instead celebrated biographically via a simple modification. For Schmidt this is linked to the biographic narrative possibilities of the built environment, as '*Architecture's greatest potential is that it's a performer. By that I mean it tells stories, produces profound connections between people and their environment*'[3] (Schmidt cited in Ingram, 2021) and MNMA clearly use materials to achieve this.

Textual Techniques

By embracing the Japanese philosophy of Kintsugi or 'golden joinery', the art of visible repair is utilised as an act of architectural redrafting, a textual technique that rephrases, modifies and punctuates the interior afresh. Aligned to the philosophy of 'Mushin', often translated as '*no mind*'[4] (Bartlett, n.d), this acceptance of the moment, of an empathetic response

Figure 5.5 By employing an 'interrupted' palimpsestic textual technique expressed as golden joinery, a site incident becomes a legible expression of the biographical life of the interior that has sustainable implications.
Copyright: Andrè Koltz.

to change that '*treats breakage and repair as part of the history of an object, rather than something to disguise*'[5] (Bartlett, n.d), was central to MNMA's creative solution. Kintsugi historically utilises lacquer or metal to repair ceramics, as the lacquer is mixed with powdered metals to create a contrasting repair seam, or alternatively uses metal staples to reset and pin broken fragments in place. As an approach, kintsugi resonates with the Japanese philosophy of Wabi Sabi in its celebration of the imperfect, the damaged and the flawed:

> Both words are difficult to translate: the former can be approximately rendered as 'poverty and undemandingness', the latter as 'seclusion, ageing, patina and decay'[6].
>
> (Iten, 2008: 18)

MNMA utilise this philosophy to express the material temporality of the site, reinforcing its biographical nature as remnants of the old combine and coexist with the new. Through separation and clarification, a site 'incident' is mapped whose emendation, or correction marks a change to the building fabric. This redrafting reinforces the notion of an 'unfinished project', a space that is evolutionary in nature as new golden veins (created by simply filling the cracks with gold dust) traverse and intersect the inherited floor surface. This simple modification interrupts and repairs the existing materiality and is integral to the ongoing life of the interior.

Sustainability

For MNMA sustainability is integral to a 'sensibility' that captures an emotional engagement with space that deepens over time. By embracing the kintsugi technique of a visible repair this material interruption patently accepts change, accidents and material degradation as integral to its life cycle. By recognising that materials have limitations MNMA embrace this potential problem as a positive source of creativity; material deterioration links to an accumulation rather than a lessening of value. This recognition leads to a simple modification or interruption that naturally embraces reuse, and extends longevity by encompassing material durability as integral to the retention of the existing building fabric. MNMA's approach to material procurement within this project helps to mitigate against unnecessary waste through material reduction, is sensitive to the circular economy and reinforces palimpsestic concerns relating to architectural redrafting such as punctuating, sequencing and editing. This ethical approach complements a desire by MNMA to embrace material memory as they believe that '*behind every material is a story* . . .'[7] (Schmidt cited in Ingram, 2021: 31).

Maison de l'Architecture, Paris, France – Chartier-Corbasson Architects

Figure 5.6 New corten steel 'patches' make legible a series of sequential modifications or 'interruptions' that punctuate the chronological historical strata that characterise this interior, as illustrated by the section drawing – Maison de l'Architecture, Paris, France by Chartier-Corbasson Architects.
Copyright: Chartier-Corbasson Architects.

Project Synopsis

This project by French architectural practice Chartier-Corbasson is an exemplary exercise in palimpsestic incidents; contemporary interruptions that have been utilised to redraft the interior via a process that 'enriches' the host building. Founded in 1999 by Thomas Corbasson and Karine Chartier, they have together developed a practice philosophy that is steeped in the history and poetics of place, have garnered numerous architectural awards and continue to marry architectural outputs with educational engagements. Many of their project proposals remain sensitive to the rich architectural heritage of France, allowing them to embrace site narrative as a constituent element of design to ensure they *'create links between the built and the existing site, between volumes, materials and lights, between the site and its history, between the program and the project . . .'*[8] (Chartier-Corbasson, n.d.).

This contextual approach actively embraces contemporary re-imaginings and additions to the historic fabric, an approach that was richly evident in Maison de l'Architecture, Paris. By capturing the essence of an interior palimpsest, expressed as a new contemporary layer, 400 years of history was carefully redrafted, a strategic approach that resulted in the 2005 European Bauwelt Prize. This precedent's inclusion in a chapter concerned with redrafting highlights the deployment of a series of palimpsestic incidents; architectural moments that 'sequence' and clarify meaning linked to temporality and narrative. Ongoing chronological incidents in the buildings tectonic and spatial organisation become a series of palimpsestic 'interruptions' that continue to punctuate, 'modify' and rephrase the interior.

Narrative

Figure 5.7 The project narrative acknowledges the legacy of this historic building by consciously extending the site biography via on site 'modifications' driven by programmatic requirements.
Copyright: Philippe Ruault.

Storytelling remains central to the site-sensitive approach of this project, an approach steeped in the building's long and chequered history from its inception as a convent at the beginning of the seventeenth century. Located in the 10th arrondissement, the Récollets convent initially housed Franciscan monks, became a military barracks after it was ransacked during the French Revolution, then a hospice, followed by a military hospital, a textile factory and even a school of architecture. The building eventually became abandoned, but not before it was briefly colonised as a squat for artists before a fire prompted their eviction. In 1999, the building was entrusted to the Régie immobilière de la Ville de Paris (RIVP), an organisation involved in social housing, urban renewal and the rehabilitation of buildings. Their involvement led to the site's current manifestation in 2004 as the Maison de l'Architecure île-de-France, a centre that aims to actively promote and disseminate discourses on the built environment via cultural debates and outreach programmes. This site biography clearly informed the design implemented by Chartier-Corbasson as their *'architectural design emphasizes the preservation of the traces of a long and turbulent history'*[9] (Maison Architecture, n.d.).

Textual Techniques

This richly-evidential historical chronology prompted an interior response that was concerned with the duality of preservation and addition, an approach aligned to palimpsestic redrafting and its accommodation of change. The retention of the dishevelled interior with its air of dereliction suggested a project solution that mediated between the past and the present, as Corbasson explains:

> First, we thought it was beautiful. Second, we had almost ho money to work with. So instead of repairing the walls and painting them white, we chose to intervene only at precise points[10].
>
> (Corbasson cited in Fayard, 2005: 264)

Sequentially this approach materialised as a series of corten steel 'patches', textual techniques that punctuate the interior and accommodate practical modifications to the historic fabric facilitated by its new functional usage. For Chartier-Corbasson these patches became *'a logical continuation of the history of the place'*[11] (n.d.), whilst the choice of a single material (corten steel) facilitated an aesthetic consistency that was itself temporal in nature. In the former barrel-vaulted chapel (now a flexible space for presentations, conferences and exhibitions), these patches shutter the windows to provide the necessary blackout conditions. Smaller vertical patches are utilised to conceal new lightning and equipment housed in existing wall bays, whilst a larger horizontal patch flips up from the floor to reveal a raised platform with step seating. These recurring, sequential motifs materialise as a contemporary colonisation of the host building, the new functional program informing the interior's ongoing historical and biographical legibility. This architectural 'punctuation' clarifies and redrafts meaning through its pacing, a process that habitually embraces change and modifies the ongoing story of the interior.

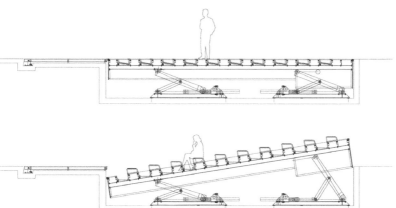

Figure 5.8 and 5.9 The largest corten steel 'patch' theatrically 'interrupts' the auditorium of the Maison de l'Architecture as it hydraulically lifts from the floor to become tiered seating (Figure 5.8). Detail drawing of the floor patch (Figure 5.9).
Copyright: Figure 5.8: Philippe Ruault. Figure 5.9: Chartier-Corbasson Architects.

Sustainability

The decision to retain much of the historic building fabric ensures sustainability is addressed at the project's inception, whilst the inherent biographical character of the interior is celebrated. Existing 'remedial' features utilised during the building's long history to prevent further collapse remain, whilst the retention of an existing mezzanine (simply resized) and staircase was driven as much by economic necessity as the site's history. The battered, scarred surfaces of the interior are redolent of its past whilst graffiti from the squatter's era endures. Inscribed words, such as 'parloir', are suggestive of its previous life as a hospital, whilst traces of ecclesiastic polychrome paint attest to the building's original incarnation. This embracing of vestigial decorative stratum, of a found aesthetic, alongside the retention of many of the remedial modifications, adds to the sense of a building that has, and is still having, a continual life. By celebrating and preserving this decorative historical stratification (in reality, sealed behind a protective layer of varnish), the building's biographical nature is celebrated. The addition of a series of sequential interruptions or modifications resulted in an economically-viable project that facilitates its new usage whilst actively entering into a debate in relation to the ongoing chronology of the building. This is a discourse that Corbasson regards as very much '*still a work in progress*'[12] (Corbasson cited in Fayard, 2005: 268). By employing architectural editing in the form of scattered, small-scale amendments, this redrafted interior is absorbed afresh into its historic fabric. This creates a legible historical reading of the context, a timeline of its use and abuse that actively recycles space and demonstrates how minimal interventions can revitalise a heritage building.

Battersea Arts Centre, London, UK – Haworth Tompkins

Figure 5.10 A listed civic building inspires a series of site-specific evolutionary palimpsestic 'adjustments' that culminates in the 'disrupted' response to the lost ceiling of the Grand Hall after a devastating fire – Battersea Arts Centre, London, UK by Haworth Tompkins.
Copyright: Fred Howarth.

Project Synopsis

If history and site narrative remain central to an interior palimpsest, then Battersea Arts Centre's evolution is a welcome addition to a chapter concerned with architectural redrafting, employed as a series of sequential amendments. Award-winning architectural practice Haworth Tompkins have been sensitive not just to the needs of the client but to the history of this former civic building located in south west London. Built in 1893 by E.W. Mountford, Battersea Town Hall is a Grade II* listed building renowned for its cultural importance in relation to women's rights via the suffragist movement, as well as its advocacy of the Labour Party and their socialist ideals. The site is composed of a municipal building to the front and a connected community space known as the Grand Hall to the rear.

After many years of service, a merging of town councils resulted in the building becoming obsolete, allowing for a change of use to its current tenants, the 'Battersea Arts Centre' (BAC), an organisation renowned as a grass roots community theatre.

Founded in 1991 by architects Graham Haworth and Steve Tompkins, 'Haworth Tompkins' are the recipients of numerous awards including the Stirling Prize. Their work with historical sites, especially the re-imagining of theatres (that began with the remodelling of the Royal Court Theatre in London), is highly regarded. As a practice, they are renowned for taking an archaeological, revelatory approach to a site's history and this modus operandi has helped them to challenge and question responses to listed buildings. This inquisitive stance always '*pay(s) close attention to the specific chemistry of individual places and cultural situations . . .*'[13] (Haworth Tompkins, n.d.). In terms of a host building's evolution and continued life, Battersea Arts Centre is an exemplary lesson in adaption that more than justifies its inclusion within this chapter. By exploiting the palimpsestic technique of redrafting, this building's trauma is reimagined as a site-responsive 'disruption' that aids reoccupation.

Narrative

Figure 5.11 and 5.12 Haworth Tompkins develop a place-based narrative inspired by the trauma of the fire. Figure 5.11 captures the devastation to the Grand Hall after the fire. Figure 5.12 highlights the lost interior that became an important source of creative inspiration.

Copyright: Figure 5.11: Haworth Tompkins. Figure 5.12: Copyright: 8Build.

By actively 'listening' to the host building and treating it as a '*memory bank*'[14] (Wright, 2019: 110) Haworth Tompkins developed a place-based narrative that ensured the past and present were involved in a continual dialogue. Consequently, the response to this historic building falls into two phases: phase two marked the reaction to a fire that partially destroyed the Grand Hall in 2015 (a momentous event that eventually became part of the site's ongoing narrative), whilst phase one incorporated all the works prior to the fire. Alongside this contextual approach, a 'scratch' working methodology, developed by the BAC Creative Director, promoted an experimental small-scale attitude to theatrical performances in an organisation that nurtures home-grown talent. By embracing the principles of scratch (essentially a work-in-progress attitude informed by collaboration and feedback), Haworth Tompkins actively 'tested' ideas in situ that questioned how they could 'reanimate' the building's neglected spaces. The impetus for this radical spatial experimentation was the staging of a performance in 2007–8 entitled 'The Masque of the Red Death' by Punchdrunk, who are leading exponents of the immersive 'promenade' approach to theatrical performance. In essence, '*Punchdrunk has pioneered a form of theatre in which roaming audiences experience epic storytelling inside sensory theatrical worlds*'[15] (Punchdrunk, n.d.). By working alongside the Battersea Arts Centre team, theatre artists and the local community, Haworth Tompkins not only helped to rethink the nature of performance but make the building 'fit' for its new purpose and functional program. Central to this ambition was the embracing of a place-based site and community-led narrative linked to the building's biography and functional usage.

Textual Techniques

By effectively becoming 'architects in residence', Haworth Tompkins, in a relationship with Battersea Arts Centre that lasted from 2006 to 2019, ensured that they were always working collaboratively via an iterative process of trial and error that utilised proposed production projects, their very staging, to make a series of sequential 'adjustments' to the building. Driven by a consolatory ethos that initially resulted in a series of scattered, phased, improvised, non-invasive alterations, these creative collaborations helped to 'unlock' the potential of the building, aiding reoccupation. These carefully-curated alterations or adjustments to the existing building aimed to ameliorate occupational difficulties by exploiting a series of design moments. This resulted in the creation of a promenade, an opportunistic arts environment where performances could take place almost anywhere.

This would have been the project culmination if it were not for a catastrophic fire in the Grand Hall and the subsequent loss of the roof (luckily, this fire did not spread to the main municipal building to the front).

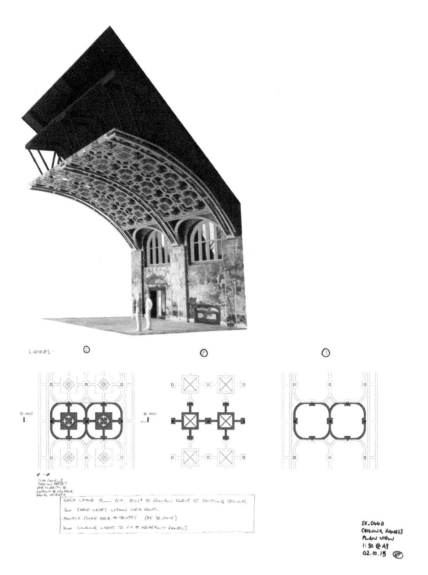

Figures 5.13 and 5.14 The model (Figure 5.13) and detail drawings of the ceiling panels (Figure 5.14) capture the palimpsestic approach, essentially a contemporary reimagining of the lost barrel vault ceiling as an architectural adjustment that redrafted but resurrected the original.
Copyright: Haworth Tompkins.

Phase two was responsive to this fire and was integral to the building's rehabilitation as a series of drawings mapped the scale of the destruction, whilst extensive documentation identified salvage opportunities. This site inventory gradually revealed a palimpsestic approach in relation to architectural 'resurrection' and 'revelation' that saw Haworth Tompkins treat the building:

> ... as a kind of phoenix rising from the ashes and we don't just build back and obliterate all memory of the fire, and we don't go back to something that existed before. We take this opportunity to think what is a 21st century grand hall like[16].
>
> (Lydon, 2021)

This realisation prompted an architectural response that was emphatically not concerned with rebuilding like-for-like, instead it created an opportunistic approach that aimed to enhance the building's biographical history, to 'redraft' the original and 'disrupt' the reality or memory of the damaged host building. The inspiration for this was the lost fibrous plaster barrel-vault ceiling and its rich Victorian ornamentation, as Lydon explains:

> Suddenly there was this sort of moment when we thought hang on what if we just focus on the plaster relief, the bit in white, what if the green solid panels become void. Suddenly, then what was once a solid ceiling becomes this lattice drape which could be hung of those roof trusses ...[17].
>
> (Lydon, 2021)

This new lattice, essentially a simplified version of the lost ceiling constructed from three layers of CNC cut plywood, followed the curvature of the original, standing in stark contrast to the fire-damaged patina of the remaining building shell below. This 'disruption' of the ceiling is a richly-evocative response to architectural memory in its 'reimagining' of this key interior attribute, allowing for improved acoustics, ventilation and the housing of technical equipment in the roof space above. This reimagined element became yet another 'adjustment' to the building, a 'disruption' that through its redrafting created further legibility in relation to the buildings chronicled history that embraced the ongoing story of this interior.

Sustainability
In 2019, Haworth Tompkins co-founded 'Architects Declare' (a network of U.K.-based architectural practices committed to addressing the climate

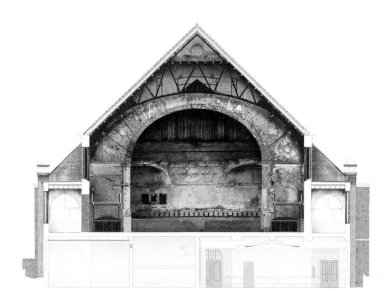

Figure 5.15 By retaining and making good the fire-damaged and water-ravaged walls, as illustrated by the section drawing, ethical decisions relating to sustainability enhance the interior.
Copyright: Haworth Tompkins.

emergency), so sustainability remains central to their practice philosophy. As a commitment to the refurbishment of existing buildings over demolition, this ethical agenda enriches the project proposal for the Battersea Arts Centre. This architectural disruption and adjustment chronicles and reveals the narrative of the town hall, a story that embraces change as part of its evolutionary 'DNA'. The site's reoccupation by a theatrical company ensured the host building had to incorporate an arts programme via adaptive reuse, a complicated functional program that was not originally intended. The employment of a transitory approach that embraced adjustments, reuse and reanimation resulted in a successful recolonisation of the building for its creative community.

The retention of the devastated interior led to a sustainable aesthetic and site response as this decorative decision chronicled the building's continual history and narrative possibilities. The fire-damaged and water-ravaged walls in the Grand Hall were carefully cleaned, stabilised and curated by a conservation team ensuring a '*conserve as found*'[18] (Lydon, 2021) approach. In contrast, the new contemporary ceiling acts as a mnemonic aid; of loss, of remembered absence as a direct consequence of the conflagration. All become integral to the sustainable incremental reinvention of a building that celebrates disruptions as architectural adjustments in its sustainable redrafting of history.

Sedgwick Rd., Seattle, Washington – Olson Kundig Architects

Figure 5.16 This heritage building successfully utilises a revisionistic 'disruption' in its reuse and repurposing of historic fragments – Sedgwick Rd., Seattle, USA by Olson Kundig Architects.
Copyright: Hans Fonk.

Project Synopsis

If interior interruptions are concerned with informing an ongoing architectural occupation, a disruption linked to a metamorphosis, then Olson Kundig's Sedgwick Rd. is a timely addition in a chapter devoted to this approach. Olson Kundig are an architectural practice based in the Pacific Northwest, founded by architects Jim Olson and Tom Kundig in Seattle (now with additional offices in New York and Chicago.). The company has acquired an impressive array of awards, industry honours and publications. Olson, as the founding partner, is renowned for his residential properties, especially art houses, that aim to exploit 'connectivity' between architecture, nature and art. Olson's guiding principle is a belief that *'architecture should fit into its context in a way that makes a better whole'*[19] (Olson Kundig, n.d.). Kundig similarly has brought a clear architectural vision to the practice, becoming celebrated '. . . *for his elemental approach to design where materials are left in the raw or natural states to evolve over time with exposure to the elements'*[20] (Olson Kundig, n.d.). His ability to design small, usually-remote cabins that utilise quirky manual mechanical devices or 'gizmo's' that are spatially transformative and deliberately tactile in nature, is integral to his oeuvre.

However, as a practice they are renowned not just for a series of considered cabins and residences, as they are involved in all aspects of the built and natural landscape, whilst their creative approach to existing buildings has to also be acknowledged. It is their attitude to adaptive reuse within the context of the Sedgwick Rd., located in Seattle, Washington, that resonates with the palimpsestic techniques of disruption. Led by Kundig for advertising firm Sedgwick Rd. (formerly known as McCann-Erickson Seattle), the project was located in the historic century-old Star Machinery Building, which the advertising agency partially leased. This project was concerned not just with the history of the building, but its possibilities as a 'found' object that could be inventoried, reimagined and repurposed for its new use and tenant. This palimpsestic response is central to its inclusion in this chapter, whilst the subsequent repurposing of the building's internal components explores design 'disruptions' that sustainably revitalise, 'adjust' and redraft the biography of the resultant interior.

Narrative

Figure 5.17 and 5.18 In situ reclamation at Sedgewick Rd. informed a project narrative of 'reanimation' inspired by Mary Shelley's infamous gothic novel *Frankenstein*.
Copyright: Tim Bies/Olson Kundig.

The project narrative is in essence two-fold: firstly driven by the client with the aim of rethinking the traditional, hierarchical nature of an office into something more open and egalitarian that enhanced creativity (surely an important attribute for any advertising agency); secondly by the desire to celebrate the age, authenticity and materiality of the building from the project's inception. Timing is everything and Olson Kundig were appointed once demolition of the existing interior was already underway, but a temporary hiatus created an opportunity for the architectural practice to take a deep breath and take stock. This pause facilitated a site inventory that led to the salvaging and storage of any reusable original interior features such as steel I beams, timber beams, doors and windows, even a crane. This strategy of reclamation informed the ongoing palimpsestic story of the scheme as the desire to repurpose and reuse, rather than discard, developed into a 'Frankenstein' narrative inspired by Mary Shelley's infamous gothic novel from 1818. The eponymous fictional character is famously composed of a series of disparate body parts that become reanimated in the creation of a new sentient being. This ethos was central to the creative approach at Sedgwick Rd.:

> Designed in the spirit of Mary Shelley's creation, where something new is created from the sum of the parts, the old pieces of the original structure were reused on moveable partition walls . . .[21].
>
> (Olson Kundig, n.d.)

This narrative of reanimation, of conscious change and sustainable adjustments, fuelled the creative response.

Textual Techniques

This literary homage resulted in a series of wheeled partition walls (quickly nicknamed Frankenstein or, more familiarly, just 'Frank') that created a creative crossroads or meeting space for the agency. By reusing existing elements for Frank, through retaining, reorganising and reimagining them, the project successfully embraced the textual technique of redrafting, of revisiting existing interior elements, to give them new purpose and meaning. These disruptions to the existing interior scenography embraced adjustments as an evolutionary aid to a continued life. This conscious repurposing, of disassembly and reassembly, is central to the nature of a disruption as '*Frank, comprises of six wheeled partitions, each a crazy quilt of multi-pane windows, peeling doors, and gnarled beams held together by*

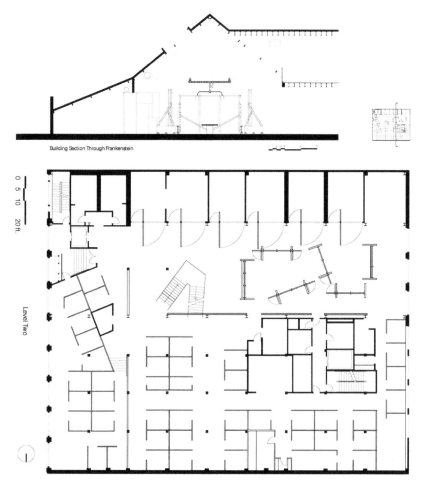

Figure 5.19 and 5.20 'Frank' became reborn by employing a 'disruptive' textual technique of reuse, disassembly and reassembly, as illustrated by the section drawing (Figure 5.19) and the plan drawing (Figure 5.20). Copyright: Olson Kundig.

new steel bracing'[22] (Kenzi, 2003: 292). This collection of found and reused elements has a hint aesthetically of theatrical staging with salvaged panels held in place by ad-hoc angled props and simple frames. This seemingly impromptu assemblage left fixings clearly on display, alongside the patina of age, of general wear and tear of the salvaged repurposed ingredients. By making these repurposed elements inherently flexible and moveable, reconfiguration of the space to suit the particular requirements of its new community was achievable.

Both light and power had to be equally mobile so they could keep pace with Frank. The solution was of course salvage, or a series of reclaimed sign lights and 1950s fluorescent louvres suspended from an overhead panel, connected to a power box with a long trailing lead. Materially, budget limitations not only benefited from a scheme that embraced the building interior as a plundered treasure trove to be reused, but also resulted in a raw, industrial aesthetic that introduced utilitarian building materials and left services exposed. As Kundig stated, '*advertising is an unfinished business, a continual process of reinvention*'[23] (Kundig cited in Kenzi, 2003: 296), so the interior echoes this ethos through its embracing of palimpsestic disrupted adjustments and biographic continuation.

Sustainability

This notion of unfinished business, of longevity, of not just invention but reinvention as a revelatory tactic, illustrates the project's sustainable credentials. Kundig believes that '*architecture is evolutionary, not revolutionary*'[24] (2019: 17), and this stance is integral to the palimpsestic reimagining or redrafting of the interior of this host building and the creative impetus behind Frank. Interestingly the alternative name for Shelley's *Frankenstein* novel is the 'modern Prometheus', Prometheus being a mythological Greek god associated with the 'creation' of humanity. As an act of co-creation with the building and its palpable history, this developmental ethos is notable; an approach that creatively reuses existing building elements. For Kundig, architecture is concerned with advancing a '*historical continuum*' as he believes that '*buildings and cities are living entities – organisms that continue to grow and evolve*'[25] (Kundig, 2019). For him, existing buildings are never 'dispensable' if we are to conserve the earth's resources and limit embodied carbon omissions. Instead, resilience, reuse and flexibility become integral to successful adaptive reuse solutions that are always conscious of their biographical nature but actively embrace palimpsestic transformation as part of a continued life. As Kundig eloquently states:

> This design strategy of reuse creates an architectural palimpsest, which has intrinsic cultural value in its layering of histories. We need to see our buildings – all of them, not just the beautiful ones – as part of a timeline, rather than a moment frozen in time[26].
>
> (Kundig, 2019)

By employing an approach that looks to both the past, the present and the future simultaneously, Sedgwick Rd., in its latest incarnation, harnesses the ethos and values of the disrupted interior. If evolution is concerned with species' adaption to an ever-changing environment, then, with the help of Olson Kundig, this interior successfully exploits these traits through its sustainable reanimation.

Military History Museum, Dresden, Germany – Studio Libeskind

Figure 5.21 In its examination of the cultural legacy of trauma and conflict, Studio Libeskind utilise 'disruptive' techniques to radically 'redraft' and 'adjust' a former armoury – Military Museum, Dresden, Germany by Studio Libeskind.
Copyright: Hufton & Crow Photography.

Project Synopsis

Studio Libeskind is an award-winnning practice that has garnered a series of high profile accolades and has an extensive back catalogue of built projects globally. Founded by architect Daniel Libeskind and his partner Nina, their work is reknowned for its engagement with emotionally-sensitive sites that often act as mnemonic aids, harnessing the poetic power of the built environment. This process prioritses a creative sensibility that often views archiecture as an act of 'reparation', that rehabilitates through direct engagement with the wider historical and cultural context. This approach is richly evident in Studio Libeskind's early commissions in Germany, namely the Felix Nussbaum House and the Jewish Museum. The former collates the works of a Jewish German Surrealist artist exterminated in Auschwitz during World War II. Nussbaum used his art to capture his experiences of Nazism and its anti-semetic polemic, whilst the latter resonates with the story of the lost Jewish community in Berlin as a result of the Holocaust. Studio Libeskind was founded in Berlin, Germany in 1989 as a direct consequence of winning the international competition for the Jewish Museum. Since then it has relocated to New York, its tranatlantic shift the result of another competition win, this time as the master planner for Ground Zero, the redevelopment of the World Trade Centre site after the terrorist attacks on the twin towers in 2001.

Given these beginnings and the resultant emotive projects, it is clear that Studio Libeskind's architectural response prioritises the regenerative potential of memory, of recollection, as opposed to collective societal amnesia. By embracing the cutural importance of remembrance, intense periods of introspection occur that not only look to the building for inspiration, but are always inclusive of its wider historical context in its understanding of place, heritage and belonging. This approach has led to a powerful association between reconciliation and 'memorialisation', as Libeskind explains:

> Well, I think without memory we would not know where we are going or who we are. So memory's not just a little sideline for architecture, it's a fundamental way to orient the mind, the emotions, the soul. And, of course, how to engage that memory through the visceral experience, not just the intellectual experience, but to the full emotional experience of a human being[27].
>
> (Libeskind cited in Deustsche Welle, 2011)

The inclusion of the Military History Museum in Dresden, Germany in a chapter concerned with redrafting illustrates the decisive power of disruptions if taken to their natural extremes. In this instance Studio Libeskind employs an architectural schism as an act of severance in an adjustment to a neoclassical building. This editorial process substantially alters the spatial and psychological meaning of the host building as a new contemporary shard is literally plunged through the building. This approach amply underlines the palimpsestic technique of disruption as a device that 'disturbs' the classical symmetry of a heritage building and drastically rewrites the chronological history of a given site.

Narrative

If an understanding of history remains central to much of Studio Libeskind's oeuvre, then this cultural resource is used to develop emotive narratives linked to place responsiveness, or design non-fiction. Often architecturally empathetic, this investigative stance highlights individuals, communities or places that have been exposed to or have suffered trauma. How this trauma is then immortalised and expressed becomes central to a redrafting, an editorial process that reworks the original host building as a form of reconciliatory emancipation. By acknowledging history and grief as an essential constituent of the human condition, Studio

Libeskind's approach to architectural adaption resonates with 'revelatory' narratives as Libeskind clearly views '*Architecture (as) a language. It has to do with history, storytelling, humanity*'[28] (Libeskind cited in Dawson, 2010: 32). Originally built as an armoury in the 1870s, the building that has subsequently become the Dresden Museum of Military History previously housed a variety of museums, all reflective of military might and propaganda that captured a shifting political polemic. The reuinification of Germany prompted a radical reappraisal, allowing the museum to be temporarily shut whilst its future was reconsidered. An international design competition invited responses to the existing historic building alongside a reconsideration of discourses connected to conflict.

Studio Libeskind were awarded the commssion with a design that challenged the competition brief restrictions of an extension that did not disturb the original neoclassical façade. For Libeskind, this form of storytelling or 'remembering' is inextricably tied to places of remembrance and reconciliation, so the redrafting of the Military Museum of Dresden is a story of flux, as he explains:

> It's a space which delibrately interrupts that chronology that ended at World War II and presents questions to a democratic society, to the republic, to families, to children: What do these conficts mean?[29].
> (Libeskind cited in Deustsche Welle, 2011)

This approach, inherently a deliberate decisive disruption of the building's biography, is reflective of a life that transforms from an aid to war, a space that promoted military propagangda, to a space that is reflective of violence.

Textual Techniques

Completed in 2011, the extension to the Military History Museum employs a textual technique of disruption that alters the existing architectural continuity and redrafts the Classical geometry of the plan, spatial expression and façade of the original grandiose host building. Often likened to a shard or a wedged arrow head, '*a massive, five storey 14,500-ton wedge of glass, concrete, and steel cuts through the former arsenal's classical order*'[30] (Studio Libeskind, 2023). This act of disruptive violence, of an architectural fracture, creates a reflective viewing platform proffering views across Dresden.

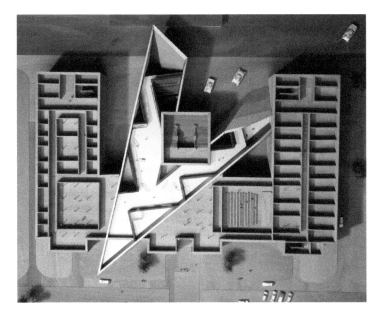

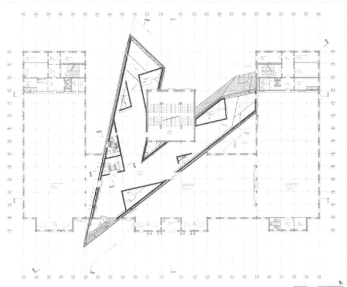

Figure 5.22 and 5.23 The model and plan ilustrate how the textual technique of 'disruption' (expressed as a wedge) fractures the existing Classical geometry and 'redrafts' the experience of the host building accommodating the Museum programme.
Copyright: Studio Libeskind.

This is a location that acknowledges, through its placement, the Allied bombardment and fire bombing of the city that led to its destuction in World War II. Additional spaces and a new sequencing arises from this disruption as the extension 'disturbs' the host building's original footprint slicing through the rear external courtyard, continuing through the building interior, to its eruptive finale on the front façade. This architectural 'disturbance' accommodates the revamped musesum's dual agenda relating to anthropological considerations of violence alongside a history of the German Armed Forces. The contemporary transparency and lightness of the extension contrasts with the existing Classical order and traditional construction methods and materials of the original. This dialetic interplay of new and old, of light and solid, is expressed through both the buiding, the thematic exhibitions and indeed Germany itself in its shift from national socialism to democractic reunification.

This palimpsestic approach embraces textual techniques that clearly edit and redraft the existing nineteenth-century building, leading to a process of architectural adaption that introduces the new by incorporating the duality of under- and overwriting. The placement, scale and expression of the new extension is deliberately disruptive in nature, illustrating the transformative shift from smaller-scale sequential interruptions or minor modifications, to assertive large-scale disruptions or more significant adjustments that together holistically underline the selection of precedents contained within this chapter. Both approaches clearly earn their place as scaled site-sensitive responses concerned with palimpsestic redrafting.

Sustainabilty

This precedent, by embracing adaptive reuse, brings a historic building back to useful life. Given its location in Dresden and its life as a former armoury (essentially a place where military arms are made and stored), alongside its connection to both military regimes and military museums, its biography was always going to be emotive. This history ensured the buildings rehabilitation was regenerative, memorialising past conflicts but acting as a place for both reflection and future discourse. This act of cultural sustainability aims to ensure that the past becomes enshrined in the present; indeed museal culture as a whole is central to the notion of temporal coexistence. This symbiosis is richly evident in the building's disruption, the collision of past and present creating a conflicted space that ressurects the existing (as well as regenerating its immediate locality as a vibrant museum district) but redrafts it symbolically. As a winner of both the Dresden International Peace Prize and the Buber-Rosenzweig medal, Daniel Libeskind is the first architect to be conferred these accolades, the latter being awarded by the Deutscher Koordinierungsrat e.V., an organisation that promotes Christian–Jewish understanding, cooperation and tolerance. Libeskind's ouevre undoubtedly views architecture as an act of cultural regeneration that actively engages in the healing power of the built environment collectively and regionally.

Chapter Five Conclusions

This chapter now seeks to conclude this investigation of palimpsestic techniques linked to **redrafting** via its exploration of '**interruptions**' and '**disruptions**'; additional techniques that provide further evidence of a palimpsestic ethos via an echoing of a theoretical stance that pervades the previous chapters. With their undoubted ability to redraft the experiential and physical conditions of a host building, interruptions and disruptions are valued as a means to reimagine, repair or rework existing content. This editorial process clearly prioritises a 'reuse' philosophy that biographically chronicles the ongoing life of a building, allowing for a considered, active evolution that mediates between minor modifications to major architectural adjustments. This recognition necessitates a reflective stance in relation to the precedents included in this chapter, a position that allows the typical characteristics and techniques of an interior palimpsest to once again be examined, and any understanding deepened. Recurrent questions reiterate concerns relating to supposed commonalities with regard to the importance of a site narrative or a building's biography, the characterisation and deployment of certain textual techniques, alongside the embracing of sustainable concerns. To help determine answers to these questions each issue will now be re-examined and conclusions drawn one last time:

Palimpsestic interiors continue to utilise narrative too:

- Embrace sequential narratives as a creative device to moderate meaning and enhance responsiveness to a specific context, place or wider historical context.
- Curate contextual stories that enhance the existing host building via new revisionary markers or 'amendments'.

Palimpsestic interiors continue to utilise textual techniques so they can:

- Integrate a sequential, evolutionary chronological progression from minor modifications or 'interruptions' to major adjustments or 'disruptions'.
- Materialise as ongoing temporal incidents or 'curatorial' moments relating to pauses or ruptures in the existing building fabric that values cultural memory and history.

Palimpsestic interiors continue to embrace sustainability because they:

- Exploit ethical recycling or rehabilitation via a legible site reading that encourages the reuse of interior components or the host building via a process that actively reduces, reuses and recycles.
- Embraces material and surface degradation, spoilt finishes and visible repairs ensuring longevity and mitigating against unnecessary waste.

REFERENCE LIST

1 Ingram, T. 2021. 'Introducing: Mariana Schmidt and Andre Pepato'. *Frame*, 138, 24–33.
2 Ingram, T. 2021. 'Introducing: Mariana Schmidt and Andre Pepato'. *Frame*, 138, 24–33.
3 Ingram, T. 2021. 'Introducing: Mariana Schmidt and Andre Pepato'. *Frame*, 138, 24–33.
4 Bartlett, C. n.d. 'Kintsugi – The art of repair'. *Traditional Kyoto* [online]. Available at: https://traditionalkyoto.com/culture/kintsugi/ [accessed 15 May 2023].
5 Bartlett, C. n.d. 'Kintsugi – The art of repair'. *Traditional Kyoto* [online]. Available at: https://traditionalkyoto.com/culture/kintsugi/ [accessed 15 May 2023].
6 Iten, C. 2008. 'Ceramics mended with lacquer: Fundamental aesthetic principles, techniques and artistic concepts'. In Christy Bartlett, James Henry-Holland and Charly Iten. *Flickwerk: An Aesthetic of Mended Japanese Ceramics*. [online exhibition catalogue]. Trans. by Howard Fine. Münster: Museum Fur Lackkunst and Ithaca, New York: Cornell University, Herbert F. Johnson Museum of Art, 18–23. Available at: https://www.raggeduniversity.co.uk/wp-content/uploads/2015/08/Flickwerk_The_Aesthetics_of_Mended_Japanese_Ceramics-ilovepdf-compressed.pdf [accessed 15 May 2023].
7 Ingram, T. 2021. 'Introducing: Mariana Schmidt and Andre Pepato'. *Frame*, 138, 24–33.
8 Chartier-Corbasson. n.d. 'Chartier+Corbasson Architects'. *Chartier-Corbasson* [online]. Available at: https://chartier-corbasson.fr/agence/ [accessed 7 June 2023].
9 Maison Architecture. n.d. 'Les Récollects, a place of history'. *Maison Architecture* [online]. Available at: http://www.maisonarchitecture-idf.org/maison-de-larchitecture/son-histoire/ [accessed 7 June 2023].
10 Fayard, J. 2005. 'Aged to perfection'. *Interior Design*, 76(7), 262–9.
11 Chartier-Corbasson. n.d. 'Architectural house'. *Chartier-Corbasson* [online]. Available at: https://chartier-corbasson.fr/maison-architecture/ [accessed 7 June 2023].
12 Fayard, J. 2005. 'Aged to perfection'. *Interior Design*, 76, 262–9.
13 Haworth Tompkins. n.d. 'Studio'. *Haworth Tompkins* [online]. Available at: https://www.haworthtompkins.com/studio [accessed 12 June 2023].
14 Wright, H. 2019. 'The walls have years'. *Blueprint*, 362, 108–24.
15 Punchdrunk. n.d. 'about-us'. *Punchdrunk* [online]. Available at: https://www.punchdrunk.com/about-us/ [accessed 29 August 2023].
16 Lydon, M. 2021. 'Architecture Project Talks: Battersea Arts Centre'. *Dezeen* [online via Zoom]. Available at: https://dezeen.zoom.us/rec/play/iVw-d3TP20qFwvcd9yXCccQRGsjC_XFuhda0gf8b6Xl5_pywThOimt5NntqvTt1QTF3wOTNfnZoykfUe.iBdLWGn3oRyuGtA_?continueMode=true [accessed 12 June 2023].

17 Lydon, M. 2021. 'Architecture Project Talks: Battersea Arts Centre'. *Dezeen* [online via Zoom]. Available at: https://dezeen.zoom.us/rec/play/iVw-d3TP20qFwvcd9yXCcc QRGsjC_XFuhda0gf8b6Xl5_pywThOimt5NntqvTt1QTF3wOTNfnZoykfUe.iBdLWGn 3oRyuGtA_?continueMode=true [accessed 12 June 2023].
18 Lydon, M. 2021. 'Architecture Project Talks: Battersea Arts Centre'. *Dezeen* [online via Zoom]. Available at: https://dezeen.zoom.us/rec/play/iVw-d3TP20qFwvcd9 yXCccQRGsjC_XFuhda0gf8b6Xl5_pywThOimt5NntqvTt1QTF3wOTNfnZoykfUe.iBd LWGn3oRyuGtA_?continueMode=true [accessed 12 June 2023].
19 Olson Kundig. n.d. 'Jim Olson'. *Olson Kundig* [online]. Available at: https://olsonkundig.com/people/jim-olson/ [accessed 15 June 2023].
20 Olson Kundig. n.d. 'Tom Kundig'. *Olson Kundig* [online]. Available at: https://olsonkundig.com/people/tom-kundig/ [accessed 15 June 2023].
21 Olson Kundig. n.d. 'Sedgewick Rd'. *Olson Kundig* [online]. Available at: https://olsonkundig.com/projects/sedgwick-rd/ [accessed 15 June 2023].
22 Kenzi, J. 2003. 'Back on Line'. *Interior Design*, 74(6), 290–97.
23 Kenzi, J. 2003. 'Back on Line'. *Interior Design*, 74(6), 290–97.
24 Kundig, T. (2019) 'Listen: Tom Kundig'. *Blueprint* 362, 17.
25 Kundig, T. (2019) 'Listen: Tom Kundig'. *Blueprint* 362, 17.
26 Kundig, T. (2019) 'Listen: Tom Kundig'. *Blueprint* 362, 17.
27 Deustche Welle; Bonn. 2011. 'Memory is essential to architecture, says Daniel Libeskind'. [online]. Available at: https://www.proquest.com/newspapers/memory-is-essential-architecture-says-daniel/docview/900315608/se-2?accountid=15894 [accessed 4 October 2023].
28 Dawson, L. 2010. 'Daniel Libeskind, Master of Memorials, on the healing power of Architecture'. *The Architectural Review*, 227(1359), 32–3.
29 Deustche Welle; Bonn. 2011. 'Memory is essential to architecture, says Daniel Libeskind'. [online]. Available at: https://www.proquest.com/newspapers/memory-is-essential-architecture-says-daniel/docview/900315608/se-2?accountid=15894 [accessed 4 October 2023].
30 Studio Libeskind. 2023. 'Military History Museum'. Studio Libeskind [online]. Available at: https://libeskind.com/work/military-history-museum/ [accessed 4 October 2023].

Interior Interruptions: A Conclusion   **Six**

Figure 6.1 Interior interruptions utilise palimpsestic techniques to embrace ethical concerns that challenge building demolition by encouraging reuse and rehabilitation.
Source: Grendelkhan, Wikimedia Commons. Licensed under the Creative Commons Attribution-Share Alike 4.0 Unported license: https://creativecommons.org/licenses/by-sa/4.0/deed.en.

OVERVIEW

If working apart we are a force powerful to destabilise our planet, surely working together we are powerful enough to save it . . . in my lifetime I've witnessed a terrible decline. In yours, you could and should witness a wonderful recovery[1].

(Attenborough, COP26 Climate Summit, 2021)

Within the context of the current climate crisis, global temperature increases, the potential collapse of supply chains and the rising cost of living, a sustainable approach that embraces the reuse of buildings has to be a priority. Worldwide interests in ethical living, environmental activism,

DOI: 10.4324/9781003326267-6

rewilding and urban ecology all link to a philosophical stance that rightly asks important questions with regard to resources, carbon footprints and waste management. Luckily, an interior palimpsest engages with these ethical issues, contributing to a rallying cry and growing mobilisation aligned to sustainable responses to the built environment. Philosophically this advantageous stance embraces 'slow' architecture, adaptive reuse and circularity by acknowledging a relationship with an existing host building that grows and deepens over time. This relational association naturally embraces the vicissitudes and vagaries of cultural memory, context and usage in the establishment of a continued rather than 'terminal' existence. This architectural continuance is typically characterised by a series of perennial interruptions, highlighting an equitable palimpsestic life within the built environment that is opportunistic in nature, responds to need and adapts to its community of users as a form of 'urban recycling'.

However, this preoccupation with the past is never wholly reverential as a palimpsestic building and its associated interior are emphatically not concerned with wholesale preservation; instead, they value a stealthier, incremental evolutionary methodology and expression. This sequential approach accommodates change through a resonance with temporal accretions and erosions, of the alterations and modifications that typically come to populate an interior palimpsest, routinely expressed as a dual aesthetic informed by a divergence of the past and the present, of the old and the new. Strategically this embraces a subtractive as well as an additive methodology to building adaption that actively utilises unwriting and overwriting, interruptions and disruptions, in the deployment of creative revisions that encourage reuse and reoccupation. This revisionary agenda shares an understanding of the built environment that resonates with place-based readings; an analysis that values architectural narratives of design non-fiction, of factually-driven site stories, that prioritises the biography or history of a building as felt phenomena. History as a continuous, systematic narrative is valued, evident in the physicality and situational condition of the rehabilitated host building. Within this context, materiality and the experiential or sensorial realm is carefully calibrated, volumetric relations reimagined, whilst the recognition of memory links cyclically back to an architectural sensibility that is redolent of cultural remembrance and place.

This recognition via a reflection of recent architectural theory acknowledges the debt of Frampton and his writings on critical regionalism, especially their interrogation of geographical and cultural context, alongside the 'peculiarities' of place as an antidote to the perceived universalism and typically raison d'être approach of Modernism (1998)[2]. A debt is also owed to Norberg-Schulz's thoughts on 'genius loci' and the phenomenology of place (1984)[3] and Zumthor's influential writings on both context and atmosphere (2015/2006)[4]. These phenomenological or situational musings all value local and regional difference, and embrace vernacular traditions alongside an inquisitorial understanding of locality or site. This sense of specificity is a preoccupation shared by the interior palimpsest, as a host building's past lives and, most importantly, their place bound narratives, become integral to a new recycled marriage of architectural convenience. This process acknowledges the building's biography, its past incarnations and is as much about retention and revelation as it is revision. The continued life, situational stories or 'legacy' of a building and its associated interior remains bound to its inherent sense of place. Scott's view that we have shifted from a Modernist sense of the ideal, highly-finished building to one that now embraces the process of ruination as '*something in process, belonging to the past, present and future, and consequently is an aspect of temporality*'[5] (2008: 96) resonates with palimpsestic thinking. The included precedents utilised to illustrate this theoretical stance strategically may vary, but all reverberate with a sense of 'incompleteness', of contributing to an ongoing <u>biographical</u> interior narrative that teases out 'revelatory' or 'revisionary' stories. These stories are concerned with a socio-cultural, as well as a situational, understanding of place and human habitation, which is inclusive of the decorative or architectural heritage of a specific building; an approach that ultimately supports an evolutionary reality. This summation resonates with the concept of an interior palimpsest as outlined in both Chapters One and Two as place-centric narratives of non-design fiction, and this inherent storytelling remains integral to the palimpsestic condition. However, this investigative journey into the preconceptions, traits, and characteristics of an interior palimpsest and adaptive reuse also highlights possible techniques that, once deployed, come to define this approach. A final, holistic appraisal of the techniques of addition and subtraction via overwriting and unwriting, interruptions and disruptions, can now be implemented for their contribution to this type of empathetic, sustainable interior.

INTERIOR INTERRUPTIONS – A SUMMATION

<u>Chapter Three's</u> analysis of the technique of '**overwriting**' explored the practice of Archiplan, Flores & Prats, Ministry of Design, TAKK and

Witherford Watson Mann via selected precedents. In this instance, overwriting is characterised by a desire to retain, recover and communicate the building's narrative as a form of layered lamination, be this decorative or more architectural in realisation. Archiplan, Ministry of Design and Flores & Prats all share an approach concerned with overwriting that, instead of obliterating the original, takes stock, assesses and creates 'pauses' that resonate with the historical, social and cultural remembrances of place. These carefully choreographed 'moments' reveal the interior's previous life, its many incarnations, successfully marrying the past to the present and accommodating temporal coexistence. In contrast, TAKK's 10K House and Witherford Watson Mann's approach to Astley Castle extend the possibilities of overwriting beyond its clear attachment to the surface to a considered 'three dimensionality'. Overwriting becomes elevational and occupational, concerned with a spatial adjacency that reveals, with an archaeological precision, a layered stratification of under- and overwriting. Both exploit the careworn or historical backdrop of the existing by introducing a new contemporary bespoke architectural fixture as a habitable residential space. Whilst the 10K House capitalises upon situational co-existence, Astley Castle goes further by embracing close contact via an architectural 'graft' that literally bonds the new to the ruined castle walls. This ensures history is ever-present in their residential settings, creating interiors that are redolent of their past, as well as of its continued, if 'interrupted', life.

Standing in opposition to the previous chapter's exploration of underwriting, Chapter Four's examination of the subtractive technique of '**unwriting**' or erasure was equally interrogated for its role in bringing a building back to an active life. Whether this was achieved through simple cuts in the existing interior surface, a repurposing of interior fixtures or a wholesale deconstruction of the building fabric, the degree of unwriting was always dictated by the existing quality of host building and its new functional programme and occupancy. This act of architectural 'modification', of partial removal and erasure, becomes a defining trait in the palimpsest's arsenal by allowing a building to be 'adjusted' as needed. This chapter's analysis of the practice of Neri&Hu, Churtichaga + Quadra-Salcedo, Heatherwick Studio, Rotor and Architecten de Vylder Vinck Taillieu identifies a commonality of methodology through the included precedents that is subtractive in nature; that exploits the unmaking, or unwriting of the existing. Neri&Hu's deployment of horizontal and vertical incisions or 'visual slippages' in the host building seeks to emulate the close urban living conditions of Shanghai via a blurring of the private and the public realms, reinterpreting this into a voyeuristic hotel experience grounded in its sense of place that recognises and celebrates ethnographic difference. Both Heatherwick Studio and Churtichaga + Quadra-Salcedo's inspired remodelling create new spatial relationships by employing a process of selective architectural erosion that archaeologically unpicks and unwrites the original, in this instance a former grain store and abandoned 1950s apartment. Both Architecten de Vylder Vinck Taillieu's response to the host building at Caritas and Rotor's reaction to a former co-operative at Mu.ZEE resist building or interior obsolesce via repurposing, employing a 'reverse' methodology expressed as industrial ecology. For the former, this subtractive process partially undoes the existing building shell in the creation of a porous, open space for all by creating a building suspended in deliberate ruination. For the latter, the interior creates opportunities for reuse that repurposes existing interior wall partitions into new spatial combinations and functional possibilities. The deployment of these revisionary techniques resists the lure of wholesale demolition; instead, they make considered deconstructive decisions that unlock new spatial, aesthetic and reclamation opportunities. This exploits the continued, if 'interrupted', life of any host building.

Finally, Chapter Five's close examination of the palimpsestic textual techniques of **'interruption'** and **'disruption'** serves to highlight how these approaches, if scaled proportionately, become a natural extension of the techniques previously discussed. By introducing an additional means for considering amendments to a host building, these ongoing temporal incidents are realised as 'curatorial' moments, essentially pauses or ruptures whose purpose is to be responsive to, clarify and reveal ongoing narratives of continued occupation. By analysing the practice through the selected precedents of MNMA, Chartier-Corbasson Architects, Haworth Tompkins, Olson Kundig Architects and Studio Libeskind, it becomes apparent that all share a congruity of purpose. Ongoing emendations, amendments and enhancements help to rephrase the existing host building, extending their respective narratives via new revisionary markers. These revisions, if scaled appropriately, can be deployed in a minimal matter, an approach shared at the Selo Shoe Store and in Maison de l'Architecure as a form of contemporaneous layering. In the former, MNMA's narrative of an inherited site problem is realised via a textual technique that 'celebrates' the necessary interruption as a visible repair or minor 'modification'. This is an approach that enhances the biographical nature of the site. In the latter, the deployment of new corten steel patches accommodates the required technical and functional constraints aligned to its new programme of habitation, whilst the resultant sequential stratification punctuates the interior and marks 'new' incidents or interruptions as ongoing chronological markers.

Both Haworth Tompkin's response to Battersea Arts Centre and Studio Libeskind's reaction to the Military Museum acknowledge the building's biography, of cultural memory and collective trauma as shared traits resulting in major disruptions. For the former, a catastrophic fire in the Grand Hall invites an architectural riposte that is reactive to this loss via an 'adjustment' that is developmental in nature. A reimagining of the original lost barrel-vault roof in a contemporaneous manner resuscitates the host building anew whilst accommodating a new technical programme. The latter's appropriation of conflict and the wider historical context of a war-ravaged Dresden inspire a contemporary disruption that, as an act of violence to the original host building, delivers a mediation on military conflict and its devastating impact. Finally, at Sedgewick Rd. Olson Kundig Architects successfully exploit the repurposed salvaged elements from the host building's previous interior occupation. This reanimation of building parts, likened to Frankenstein's monster, marks an alteration that physically disrupts and redrafts the original. This concise summation reveals a distinctive architectural nuance oscillating as it does between interruptions and disruptions, from minor modifications to major, or more significant physical adjustments. What becomes patently obvious is that this is about a legible site reading that actively reduces, reuses and recycles as a constituent aspect of any architectural redrafting. This 'reappraisal' is clearly concerned with a rewording of the original, a translation that actively creates new meaning, that accommodates change, embraces interior interruptions and values the continued life of a building by always paying close attention to the 'chemistry' of any given place.

From a sustainable perspective, it is evident that palimpsestic interiors prioritise adaptive reuse and the repurposing of existing buildings, an approach that helps to 'minimise' building waste as the host site is instead assessed in terms of its longevity and reuse potential. This stance resonates with and acknowledges the 'whole life performance' of a building, of its continued circularity in relation to its life cycle. By recognising the value of an existing building as a 'found' object, its inherent personality, cultural resonance and memory become important drivers of design invention that values adaption as a creative, sustainable starting point. As the construction industry remains one of the main contributors to global carbon emissions, the decision to 'extend' rather than 'end' the life of a building is an important one sustainably. By utilising site investigative techniques to access the value and possibility of reuse, an introversion is encouraged that prioritises the recycling of existing materials and interior scenography.

This archaeological approach to the built environment embraces an investigative stance which treats the host building as a valuable resource that is akin to an autopsy as it peels back layers, exposes scars, makes incisions, but always rehabilitates through remedial work. These unique contextual atmospheres inform the interior palimpsest as the host building is literally 'mined' for creative opportunities. In its furtherance of the circular economy and reduction of resource consumption, a 'found' aesthetic, an unfinished, even 'spoilt', language that opportunistically integrates architectural salvage, abandoned domestic and urban detritus, is typically ubiquitous. By embracing this approach, all are mindful of the CO_2 emissions or the embodied carbon associated with the lifecycle of a building. Additionally, a significant percentage of the case studies explore smaller habitable spaces within a larger building footprint as a means to reduce energy consumption and mitigate against operational carbon. By retaining more of the existing fabric, by resisting the urge to remove, replace or simply encase, waste and material consumption is reduced, and embodied carbon emissions minimised. The retention of virgin materials and the recognition that an interior aesthetic does not need to be about surface perfection liberates cyclical opportunities for reuse that embraces responsible material sourcing, consumption and negates building obsolescence.

Whilst some buildings go into a forced dormancy or temporary abeyance because they are surplus to the requirements of industry or their inhabitants, or because financially they are no longer viable, others experience a savage rupture in their timeline caused by human or natural catastrophes. Alternatively, some retain a peaceful, equitable existence. Waking these buildings up and bringing them back to life, acknowledging their past biographies and allowing their previous lives to 'taint' or 'contaminate' any new incarnation is central to the creation of an interior palimpsest. It would be naïve to suggest, given the Anthropocene age we are currently experiencing, that palimpsestic interiors are the answer to global environmental concerns. However, as a considered response to an ever-industrialised world, to humanities predilection for mass consumption, it proffers a possible antidote that when combined with other environmental measures can contribute in a meaningful manner.

This chapter, alongside the previous chapters, has aimed to highlight the theoretical concerns, characteristics and archetypal techniques common to an interior palimpsest. This palimpsestic emphasis is important because it recognises the revisionary capabilities and editorial capacities of building reuse, celebrates its innate biographical narrative, prioritises a sustainable agenda and champions textual techniques with their innate ability to reuse, redraft and rewrite content. This revelation rightly acknowledges palimpsestic interiors and their congenital, hereditary ability to recycle, reuse and rehabilitate. This recognition serves as an argument for the wholesale consideration of '**interior interruptions**' given their capacity to engender the sustainable rehabilitation of the old in the creation of the new. Buildings and their interiors no longer need to be fixated by the new; instead, they can become a literal 'accretion' of the life they have lived and continue to live.

REFERENCE LIST

1 BBC News. 2021. 'COP26: David Attenborough says world is looking to leaders'. *BBC News* 1 November [online]. Available at: https://www.bbc.co.uk/news/world-59125138 [accessed 13 January 2024].
2 Frampton, K. 1998. 'Towards a Critical Regionalism: Six Points for an Architecture of Resistance'. In Hal Foster (ed.). *The Anti-Aesthetic: Essays on Postmodern Culture*. New York: The New Press.
3 Norberg-Schulz, C. 1984. *Genius Loci: Towards a Phenomenology of Architecture*. (1980). New York: Rizzoli.
4 Zumthor, P. 2006. *Atmospheres: Architectural Environments, Surrounding Objects*. Reprint 2015. Basel: Birkhäuser.
5 Scott, F. 2008. *On Altering Architecture*. London and New York: Routledge.

Index

10K House (*see also* TAKK) 55, 72–77, 162; CNC technology, usage 76–77; development 73; exterior house, plan perimeter 75–76; financial constraints 73f; heat, emission 76; host building sensitivity 73–74; interior fixture/insulated house 76f; material consumption, reduction 73f; materials, reductionist approach 75; natural cross ventilation 75–76; original floor plan, reminder 74; palimpsestic textual techniques 74f; palimpsestic traits 72f; project narrative 73–74; project synopsis 72–73; revisionary temporal narrative 74; sustainability 75–77; textual techniques 74–75

168 Upper Street (Taha+Groupwork) 12, 14–15

adaptability, implication 7
adaptation, term (usefulness) 6
adaptive reuse 4, 8; colonisation 18; debate 2; discipline 52; practice, understanding 3; process 148; term, usefulness 6; value 39–40
addition/subtraction 54; process 32–33; usage 53
additive methodology 160
additive process, impact 88
additive techniques 54, 124
additive textual techniques (palimpsestic interior) 51f
advertising, reinvention 148
Affiches lacerees 33
Alexander Nevsky montage (Eisenstein) 37f

alteration 6; consequence 42; process 41; usage 53
alternative vision, offering 37
ambience, result 105
American Society of Interior Designers (ASID), design excellence (focus) 9–10
anarchitecture 34
ancient manuscripts, damage (Middle Ages) 30
Anthropocene Age 7, 165–166
archaeological dig, physical act 31
Archiplan Studio (*see also* Broletto Uno Apartment) 55, 56–57, 59, 161, 162
Architecten de Vylder Vinck Taillieu (a DVVT) 89, 114, 118, 162–163; interest 115–116
'Architects Declare': Haworth Tompkins, co-founding 141–142; pledge 103
architectural adaption, Studio Libeskind approach 152
architectural appropriation, problems 8
architectural change, accommodation 46–47
architectural continuance 160
architectural creation, conceptual instrument 42–43
architectural disruption, ch+qs interest 98
architectural diversity 107
architectural editing 115
architectural errors/mistakes, inclusion 15
architectural fragments, reuse/relocation 7
architectural ghosts 71
architectural legacy 8
architectural modifications/alterations 39–40, 54–55

architectural narrative, building biography (relationship) 42
architectural pruning 110–111
architectural punctuation, impact 134
architectural rebirth, salvage (usage) 71
architectural redrafting 124
architectural remembrance, seminal example 65
architectural remodeling, interpretation 41
architectural resurrection 141; process 118–119
architectural resuscitation 66
architectural revelation 141
architectural shell, viewpoint 3
architectural sites, exploration 41
architectural storytelling 42; form, creation 23
architecture: amelioration 124; evolution 148; nature/art, connectivity 144–145
artistic practice: alignment 17; palimpsestic methodologies, parallels 33
art palimpsest 31–34, 32f
Astley Castle (see also Witherford Watson Mann Architects) 55, 78–83, 162; architectural graft, full contact 82; brick clasps, usage 82; building history, consideration 79; evolution, models 82f; historical layering, visual cohesion 82–83; historical resonance 79; historic ruin/proposed dwelling, biographic relationship 81; host building, chronological biography 78f; host building, surface treatment 82–83; incompleteness, narrative (reinforcement) 83; narrative 79, 81; national significance 79; project narrative, half house/half ruin 80f; project synopsis 78–79; rebirth 79; sustainability 83; textual techniques 81–83
Austen, Jane 35

"Ballad of Julie Cope, The" (poem) 14
Banksy 12, 16–17
barrel vault ceiling, reimagining 140f, 164
Battersea Arts Centre (see also Haworth Tompkins) 123f, 126, 137–143; BAC tenants 138; barrel vault ceiling, reimagining 140, 164; ceiling disruption 141; ceiling panels 140; CNC cut plywood, usage 141; consolatory ethos 139; detail drawings 140f; devastated interior, retention 142–143; history, archaeological revelatory approach 138; lost interior, creative inspiration 138f; memory bank, treatment 138–139; narrative 138–139; non-invasive alterations 139; place-based narrative, usage 138f; project proposal 142; project synopsis 137–138; response 164; sustainability 141–143; textual techniques 139–141; walls, retention/improvement 142f
Belgian Building Research Institute 108
Belgium, Surrealist heritage 114–115
Benjamin, Walter 18
Biblioteca Capitolare 30
'Bingo' (Matta-Clark, Gordon) 34
biographical interior narrative 161
biographical narrative 125; innate biographical narrative 166
biographical timeline, revealing 112–113
biographic contextual storytelling, enhancement 58f
biography, notion (extension) 42
Bombay Sapphire refurbishment project, Outstanding BREEAM rating award 103
Botton, Alain de 14
brick clasps, usage 82
bricolage: decorative bricolage 69; embracing 65
British Institute of Interior Design (BIID): Sustaining Specifying Guide, carbon footprint report 2; UNSDGs (embracing) 9
Broletto Uno Apartment (see also Archiplan Studio) 51f, 55, 56f; inclusion 57; project narrative 57–58, 61–62; project

synopsis 56–57, 60–61; sustainability 59, 64; textual techniques 58–59
Bronte, Emily 35
Brooker, Graeme 43
building: adaption, prioritisation 8; assets, recognition/valuation 4; biographic past 62; concrete building, erosion 102f; creation 163; cultural artefacts, celebration 8; dialogue, development 43–44; dissections (Matta-Clark) 98–99; envelope 3; fabric, retention 131; historical evolution, attention 59; historical resonance 58; history, prioritisation 160; host building 78f, 90; host, reference 3–4; innate beauty 56–57; interruption 42–43; legacy 161; life cycle/intrinsic value, importance 42; longevity, perception 9; materials, reuse/relocation 7; obsolescence, Mu.ZEE resistance 163; ongoing legacy 120; remodelling 43; renovation 63–64; repositioning 2; repurposing 78–79, 164–165; upgrading 6; waste, minimisation 120
building biography 57; acknowledgement 47; affiliation 52; architectural narrative, relationship 42; commonality 88; prioritisation 160
Building Research Establishment Environmental Assessment Method (BREEAM) 10
building reuse: capacities 166; encouragement 10; palimpsest, relationship 1–2
built environment: archaeological approach 165; biographic narrative possibilities 129; intangible aspects 96–97; palimpsestic reading, establishment 10; palimpsestic understanding, reinforcement 43; physicality 97; storytelling, categories 12
built interiors 4

Canvas House (see also Ministry of Design Pte ltd) 55, 60f; project narrative 61–62; placed-based temporal narrative 61f; placed-driven narrative 62; project synopsis 60–61, 65–67; renovation 61; surfaces/materials/fittings/furniture 63f; sustainability 64; textual techniques 62–64, 67–71
carbon dioxide (CO_2), embodied carbon 8–9
carbon emissions 165; minimisation 118–119
carbon footprint, construction industry account 2
Caritas Psychiatric Centre (see also Architecten de Vylder Vinck Taillieu) 89, 114–119; architectural editing 115; architectural resurrection, process 118–119; building, retention 116; deconstruction, strategy 114f; greenhouses, introduction 117–118; incidental occupational moments 116; narrative 115–118; notional garden rooms, presence 117–118; opportunistic approach (sketches) 116f; partial demolition 118; project synopsis 114–115; protection/comfort, creation 117–118; remains, stabilisation 117–118; site evolution, exploitation 118; subtractive approach 118; sustainability 118–119; unmaking (palimpsestic textual technique) 117f; user community, physiological/psychological needs 119
Casares, Toni 67
case studies: analysis 52; selection, deliberateness 88–89
Castelvecchio Museum, interior palimpsest example 40f
ceilings, removal 112
celluloid palimpsests, city chronicling 39
ceramics, repair 130
change, accommodation 10
Chapter of Verona library 30
characteristics, investigation 5

Index

Chartier-Corbasson Architects (*see also* Maison de l'Architecture) 126, 132–136; design implementation 134; precedents, practice analysis 163–164
Chartier, Karine 132
Chipperfield, David 45
chronological episodes 41–42
chronological history, rewriting 151
Churtichaga, Josemaría de 96, 99
Churtichaga + Quadra-Salcedo (ch+qs) (*see also* HUB Flat) 162–163
cinematic palimpsests 37f, 38; inclusion 39
circular economy 7; embracing 113; link 9
Circular Economy Package, European Parliament adoption 10
circularity, integration 10
climate (design excellence tenet) 9–10
climate crisis 159; resolution 103
close layering 53–54
CNC cut plywood, usage 141
CNC technology, usage 76–77
Coal Drops Yard (Heatherwick Studio) 103
Codex Ephraemi Rescriptus, location 30
Codex Guelferbytanus 64 Weissenburgensis (text) 28f
Codex rescriptus, term (preference) 29
Coetzee, Mark 104
coexistence, norm 7
collage/décollage, artistic practice 32–33
collective social memories 67
collective space, gradients 118
comfort/discomfort, voyeuristic interplay 93
commonalities: establishment 34–35; investigation 5
common materials, extraction/reintegration 109
comparative parameters, series (creation) 29
concrete building, erosion 102f
concrete silos, architectural erosion 103
connectivity, exploitation 144–145

Connor, David (*see also* Croft Lodge Studio) 18, 21–24
conscious repurposing 146–147
construction methods 97
contextual atmospheres, uniqueness 120
contextually-driven site-specific meaning 125
contextual stories, curation 156
continual resurrection/reuse/re-inscription cycle, creation 46
continued life 2
conversion, conceptual instrument 42–43
cooperativa pz y Justicia 55 (*see also* Flores & Prats)
cooperative, temporal timeline 66
Corbasson, Thomas 132, 134, 136
corrosion/contamination 46
corten steel: material, choice 134; patches, usage 132f, 135f
creation, emphasis 15
creative disciplines, impacts 5, 31
creative revisions 4
creative translation, act 54
creativity, positive source 131
Croft Lodge Studio (*see also* Kate Darby Architects/David Connor) 18, 21–24, 22f; plans 23f; work, precommencement 23f
Crystal Houses (MVRDV) 18, 19–20, 24
cultural context, interrogation 160–161
cultural ghosts 71
cultural memory, vicissitudes/vagaries 160
curatorial moments, impact 156
Cuypers, Pierre 19
cyclical model, development 7

damaged, celebration 130
David Chipperfield Architects, Neues Museum remodelling 45
David Connor Design 21, 24
Deconstruction and Reuse: How to Circulate Building Elements (Rotor/Belgian Building Research Institute) 108
deconstruction, strategy (example) 114f

deconstructivist literary theory 36
decorative blank canvas 64
decorative bricolage 69
decorative historical stratification, celebration/preservation 136
decorative temporality, conscious design choice 21
degradation/attrition 46
democratic reunification 154
De Quincey, Thomas 35
Derrida, Jacques 36
design excellence, ASID focus 9–10
design fiction 12; examination 19–20; fictional site narratives, contrast 12–17
design non-fiction 12, 52; exploration 18; factual site narratives, contrast 18–24; place-bound revelatory narrative 57–58
Deutscher Koordinierungsrat e.V. awards 155
developmental strategy 59
Dickinson, Emily 35
digital inaccuracies 15
Dillon, S. 35
disassembly/reassembly 146–147
disruptions: ability 36; examination 124, 156; palimpsestic technique 151; palimpsestic textual techniques 163; techniques, deployment 6
disruptive violence, impact 152, 154
diurnal patterns 18
dust, celebration 23–24
dystopian colonial theme 16

economic viability, ensuring 64
Eisenstein, Sergei 39
Elena Almagro (Churtichaga+Quadra-Salcedo): palimpsestic interior 87f
embodied carbon, operational carbon (contrast) 8–9
emotional engagement, exploitation 107
Encyclopedia of Ancient Christianity (Milazzo) 29
Encyclopedia of Ancient Literature (Cook) 29
end-of-life 2
entrance screens (architectural feature) 64
environmental awareness 72

Ephraem the Syrian 30
equity (design excellence tenet) 9–10
essentialism 61
ethical production/consumption, cyclical model 8–9
European Parliament, Circular Economy Package adoption 10
evolutionary adjustments 126
experiential atmospheres, ch+qs interest 98
experiential narrative 129
experiential realm 160
expressive site narratives, celebration 3
extraction, prioritisation 7
extreme conservation 21
Eysselinck, Gaston 110

facsimile 15
factual, fictive (contrast) 11–12
factual information, history concerns 38
factually-driven site: responses, literary appraisal 18; stories 160
factual narrative: capture 21; uncovering 19–20
factual site narratives 22f; design non-fiction, contrast 18–24
Fashion, Architecture, Taste (FAT) 12–13
Federico II Gonzaga (Duke of Mantua) 57
Felix Nussbaum House (Studio Libeskind) 150
feminist rereading 36
fictionalised misremembering 19–20
fictional site narratives, design fiction (contrast) 12–17
fictional subterfuge 15
fictitious site, responses 17
film essayist: milieu 37; work, understanding 39
filmic techniques, inclusion 38–39
flâneur, concept 37, 39
flawed, celebration 130
Flemish Belle Epoque-style buildings/pavilions 115
flexibility, implication 7
'Flora' (Rembrandt) 32

Flores & Prats (*see also* Sala Beckett) 55, 65f, 161, 162; building veneration 66–67; models/drawings, production 67; palimpsestic textual techniques 68f; rehabilitation 66–67
Flores, Ricardo 65
found aesthetic, retention/enhancement 120
found object 164–165; possibilities 145
'Found' recycled aesthetic 46–47
Frankenstein (Shelley): inspiration 145f; literary homage 146; monster 164; name, alternative 148; narrative, development 146
fresco remnants, preservation 58–59
fugues 98–99; control 99f

gender equality, questioning 35–36
genius loci 3
geographical context, interrogation 160–161
geological epoch, defining 7
German Armed Forces, history (impact) 154
Germany, reunification 152
Gesamtkunstwerk (total work of art) 14
'Ghost' (Whiteread) 34
ghostly biographical remnants 66
ghosts: collective social memories 67; re-creation 23
Gilbert, Sandra 35
glass bricks, development 20
globalisation, increase 19
global temperature, increase 159
gold dust, usage 131
golden joinery. *See* Kintsugi
graffiti, street art 33
Grain Silo complex, reuse 104
Graphein, meaning 33
greenhouse gas (GHG) emissions, embodied carbon 8–9
greenhouses, introduction 117–118
green wayfinding 69
Ground Zero, plan 150

groupwork 12, 14–15; design, result 15
Gubar, Susan 35

habitable space, creation 79
Hage, Rabih 46
Hains, Raymond 33
half house/half ruin 80f
Haworth, Graham 138
Haworth Tompkins (*see also* Battersea Arts Centre) 126, 137–143; archaeological approach 45–46; Battersea Arts Centre (BAC), relationship 139; precedents, practice analysis 163–164; sensitivity 137
health (design excellence tenet) 9–10
Heatherwick Studio 89, 162–163; appointment 104–105; credentials 107; ethical agenda 103; founding 102–103; global reach 103
Heatherwick, Thomas 102–103
Heidegger, Martin 36
Historia philothea (Theodoret of Cyrrhus) 30
historical accuracy, reliance 19–20
historical buildings, rehabilitation 42–43
historical interior vignettes, descriptions 18
historical layering, visual cohesion 82–83
historical legacy 8
historical memory 38
historic building: deconstruction 115; fabric, retention decision 135–136
historic interior surface treatments, patchwork 68–69
historic layers, notions 46
holistic disruption 124
Holland, Charles 12–13
homogeneity, increase 19
host building: chronological biography 78f; palimpsestic approach 124; response 94; spatial/psychological meaning, alteration 151
House for Essex, A (Perry/Holland) 12–14, 13f
HUB Flat (*see also* Churtichaga + Quadra-Salcedo) (ch+qs) 89, 96–101; narrative

97–98; palimpsestic approach 100f; palimpsestic interior 96f; palimpsestic textual techniques, deployment 99f; polytechnic senses 96–97; project narrative 97f; project synopsis 96–97; reductionist approach 98; renovation 97; space, transformation 97f; sustainability 100–102; textual techniques 98–100; transfusions 96
humbleness, implication 7
Hu, Rossana 90
Huyssen, Andreas 38, 39

iconic Victorian female authors, Feminist rereading 36
immigrant mercantile communities, building construction 62
imperfections, celebration 21, 130
incompleteness: narrative 81; notion 43
indigenous archetype 92
industrial buildings, rehabilitation 42–43
infra-ordinary 18
innate biographical narrative 166
Inside Quality Design (IQD) 65
in situ reclamation (Sedgwick Road) 145f
in-situ testing, usage 112
installation/intervention/insertion, terminology (usage) 43
Institutes of Gaius 30
inter-generational living, norm 7
interior(s): biography/historiography 34; historiography 31; landscape, reimagining 98–99; mining 100–101; narrative, redrafting/rewording (translational techniques) 54–55; obsolescence, Mu.ZEE resistance 163; revisionary life 88; textual analogy 53–54
interior architecture: discipline 52; value 39–40
interior design, discipline 52
interior interruptions 1–27 (chapter);124, 161; consideration 166; exploration 5; palimpsestic techniques,
usage 159f; series 3; synopsis 1–7; understanding 52
interior palimpsest 39–47, 61; advantage 53–54; aspects 6; Castelvecchio Museum example 40f; characteristics 5; concerns/preoccupations 6, 128; creation 165; engagement 160; initial definition, expansion 52; populating 160
interior sphere, presence (divulging) 41
interior story, response 57
interrupted interior, palimpsest (relationship) 1–2
interrupted palimpsestic textual technique, usage 130f
interruptions 8. *See also* Interior interruptions; examination 124, 156; palimpsestic textual techniques 163; relationship 43; series 4; techniques, deployment 6; type 54, 88, 125–126

Jewish Museum (Studio Libeskind) 150
Julian Harrap Architects 45

Kate Darby Architects (*see also* Croft Lodge Studio) 18, 21–24
kintsugi (golden joinery) 128; embracing 129–130; expression 130f; ideals 46; lacquer/metal, usage 130
Kolumba Art Museum 44f
Kopytoff, Igor 42
Kuleshov, Lev 39
Kundig, Tom 144, 148, 149

Landmark Trust 78–79; proposed brief 79, 81
lane house, story translation 92–93
late-Gothic Kolumba church, historic remnants 44, 44f
Leadership in Energy and Environmental Design (LEED) 10
Libeskind, Daniel (awards) 155
literary palimpsest 34–36, 34f
literary theory, pillaging 36

literary tropes, impact 35–36
Living Architecture, holiday rental 14
living organisms, reproduction/metabolism 3–4
London Plan (Khan) 10
Luzárraga, Mireia 72

MVRDV 18, 19–20, 24
machine milled polystyrene moulds, construction 15
Maison de l'Architecture (see also Chartier-Corbasson Architects) 126, 132–136; architectural punctuation, impact 134; auditorium, interruption 135f; building legacy, project narrative acknowledgment 133f; context, historical reading 136; corten steel patches, usage 132f, 135f; decorative historical stratification, celebration/preservation 136; historic building fabric, retention decision 135–136; narrative 133–134; place, history (logical continuation) 134; project synopsis 132–133; RIVP, impact 133–134; site biography, extension 133f; site-sensitive approach, storytelling (impact/importance) 133–134; small-scale amendments 136; sustainability 135–136; textual techniques 134
manuscript, creation/revision (parallels) 41–42
'Masque of the Red Death, The' (Punchdrunk staging) 139
material: deterioration 131; durability 9, 131; reduction 131; reductionist approach 75; repurposing 113; resilience/efficiency 9; reuse, encouragement 10
material-led project solution 128–129
Matta-Clark, Gordon 33–34, 98
McCann-Erickson 145
meaning: accumulation 36; fluidity 48
memorialisation, reconciliation (association) 151

memory: bank, treatment 138–139; collective social memories 67; cultural memory, vicissitudes/vagaries 160; historical memory 38; links, recognition 160; playing 15; repositories 61
Ménard Dworkind Architecture & Design (MRDK) (see also RYÙ Peel) 18, 20–21, 24
Merzbau (transformative process) 33
Merz Pictures, formation 32–33
Military History Museum (see also Studio Libeskind) 126, 150–155; classical geometry/redrafts 153f; disruptions (decisive power), redrafting (impact) 151; disruption, textual techniques (model/plan) 153f; disruptive techniques, usage 150f; extension, completion 152–153; host building, spatial/psychological meaning (alteration) 151; narrative 151–152; neoclassical façade, disturbance (absence) 152; project synopsis 150–151; reconciliatory emancipation 151–152; redrafting 152; response 164; sustainability 154–155; textual techniques 152
mimicry, act 15
Ministry of Design (MOD) 161, 162; design philosophy 64; internal surface coverage 62–63; plaudits/awards (see also Canvas House) 60
Ministry of Design Pte ltd 55, 60f
mirrored window shutters 94
MNMA (see also Selo Shoe store) 126, 127–131; kintsugi, usage 130–131; precedents, practice analysis 163–164; sustainability 131
modernism, approach 161
montage, filmic device 39
Mountford, E.W. 137
Muiño, Alejandro 72
multi-layered palimpsest 37
Mushin (no mind) philosophy 129–130

Mu.ZEE Art Museum Redesign (*see also* Rotor) 89, 108f; formation 110; narrative 110; site history, acknowledgment 112–113; sustainable process, development (example) 112f

narrating process 3
narrative: abandoned grain store/silo, narrative 104f; acknowledgement 47; biographical narrative 125; experiential narrative 129; innate biographical narrative 166; invention 17; misremembering 14–15; Mu.Zee narrative 110; palimpsest, relationship 11; place-bound narratives 120; place-centric narratives, defining 18; response 109f; space, interest 18; storytelling, examination 52; UNESCO World Heritage Site 57–59; urban narratives, changes 39; usage 11–12, 120
national socialism, shift 154
Nazism, experiences (capture) 150
negative spaces, concrete cast 34
neoclassical façade, disturbance (absence) 152
neoclassical street façade, disruption 15
Neri&Hu Design and Research Office (*see also* The Waterhouse at South Bund) 89, 90–95, 162–163; nostalgia, embracing 92
Neri, Lyndon 90
net zero carbon building, defining 10
net zero whole life carbon, RIBA belief 2
Neues Museum, architectural remodelling 45f
New Majestic Hotel, MOD commission 60
nong tang, story translation 92–93
Nouveaux Réalistes, work 33

Of Grammatology (Derrida) 36
old buildings, listening (advantages) 18
'Old Guitarist' (Picasso) 32
Olson, Jim 144

Olson Kundig Architects (*see also* Sedgewick Rd.) 126, 144–149; precedents, practice analysis 163–164
operational carbon, embodied carbon (contrast) 8–9
oppositional subtractive preoccupation 124
original: reconsideration/reorganisation/reimagining 124; redrafting 141; resource, viewpoint 47
Ostend's Museum for Fine Arts (OMSK) 110
outstanding BREEAM rating, award 103
overpainting, practice 31–32
overwriting: 51–86 (chapter); evidential change 30–31; palimpsestic devices/techniques 53–54; process, impact 41–42; rewriting/unwriting, conflation 124; technique 6, 161–162; unwriting, combination 89

palimpsest: 28–50 (chapter); ambition, connection 2; art palimpsest 31–34, 32f; basis 24; built environment/building life, parallels 3; celluloid palimpsests, city chronicling 39; cinematic interpretations, alternatives 38–39; cinematic palimpsests 37f, 38; definition 29–31; establishment 39–40; example 28f; globally-significant examples 30; identification 34–35; impact 36; importance, reiteration 38; interdisciplinary analysis 29; interior palimpsest 5–6, 39–47; involuted phenomenon 35; literary palimpsest 34–36, 34f; multi-layered palimpsest 37; narrative, relationship 11; nature, clarification 5; notion, reinforcement 46; preconceptions 29–30; redraft, ability 6; role, examination 1–2; sustainability, relationship 1–2, 7–10; theoretical role, examination 5–6; traits, establishment 31; types/differences 31–48

palimpsest characteristics: establishment 31; investigation 5
palimpsest commonalities 31–48; investigation 5
palimpsestic additive technique 54
palimpsestic interior 96f; narrative, usage 120, 156; repositioning 2; sustainability, link 120
palimpsestic manuscript, integrity 39–40
palimpsestic methodology 3; artistic practice, parallels 33
palimpsestic nature 66
palimpsestic process, impact 46–47
palimpsestic reading, establishment 10
palimpsestic redrafting 134
palimpsestic re-imagining 5
palimpsestic techniques 87–122 (chapter); 6; deployment 125
palimpsestic textual techniques, deployment 58f, 99f
palimpsestic thinking: investigation 31; pertinence/prevalence, identification 52
parloir 136
partial dismantling, process 41–42
partial erosion, revealing 106f
partially-demolished building, deconstruction 89
partial unwriting, subtractive textual technique 90f
past: dogmatic reverence 4; preservation 98
Peace and Justice Cooperative building, renovation 65
peace/prosperity, shared blueprint 9
peek-a-boo approach 62–63
pentimento, artistic practice 31–32
Pepato, André 127
Perec, Georges 18
Perry, Grayson 12–13
personal space, visual/aural/physical limitations 93
physical deconstruction 88–89
physical stratification 53–54
Piazza Broletto 57
place: activation 43–44; examination 90–91; genius loci 3; history, logical continuation 134; phenomenology 161

place-based heritage, examination 90–91
place-based readings 160
place-based sensitivity 91
place-based temporal narrative 61f
place-bound narratives 120
place-bound revelatory narrative 57–58
place-centric narratives 161; defining 18
place-centric revelatory story 21
place-driven narrative, delivery 62
polytechnic senses 96–97
postmodern, oppositional view 38
Prado Museum 97
Prats, Eva 65, 67
precedent selection, deliberateness 125
present futures 38
present pasts, concept 38
produce-use-discard, prioritisation 7
project narrative: delivery 128f; two-fold characteristic 146
Provincial Museum for Modern Art (PMMK) 110
Punchdrunk (theatre) 139

Quadra-Salcedo del la, Cayetana 96

reanimation: narrative, impact 146; project narrative (Sedgewick Rd.) 145f
recognition, process 41–42
reconciliation, memorialisation (association) 151
re-creation, impact 15
recycling, issues 43
redraft, ability 6
redrafting 123–158 (chapter), 125–126, 156; act 54; architectural redrafting 124; process 41, 124; textual technique 146–147
reduce, reuse, recycle (mantra) 7
Refined Rococo Townhouse Style, building dates 62
refurbishment, catalyst 21
Régie immobilière de la Ville de Paris (RIVP), impact 133–134
regional vernacular typologies 92; communal domesticity 91
reinforced concrete, skeleton 110–111

reinterpretation: importance 46–47; possibility 36
reinvention, act 54
reinvigorating/reviving, act 6–7
relational association 160
remedial work, visible mending 46
remodelling: inspiration 163; perspective 41
Renewed old, classification 70
renovation, term (usage) 6–7
Rereadings (Brooker/Stone) 43
resource efficiency 7
Restoration, curative nature 6–7
restrictive geometry, unlocking 105
retail space, global brands (presence) 19
retention 6; impact 59; importance 46–47; usage 53
'Retro First' (campaign) 103
retrofitting, term (usage) 6
reuse 111; capacities 166; circularity 2; design strategy 148; history, link 4; notion, importance 30–31
'Reuse Toolkit' 109
re-utilization, term (usage) 42–43
revelation: act 54; importance 46–47
revelatory stories 161
revenants, embracing 42
Reverse Architecture (Devlieger) 108
reverse methodology, usage 108f
revisionary stories 161
revisionary techniques, usage 120
revisions, embracing 42
rewriting: overwriting/unwriting, conflation 124; palimpsestic devices/techniques 53–54; palimpsestic methodology 3
risky reading 35
Romano, Giulio 57
room, surface treatment 39–40
Rotor (collective) 89, 108, 162–163; narrative response 109f; selective dismantling (*see also* Mu.ZEE Art Museum redesign) 109–111, 111f
Rotor Deconstruction and Consulting (Rotor DC) 108–109
Rough Luxe Movement 46

Royal Court Theatre, remodelling 138
Royal Institute of British Architects (RIBA): award 78–79; net zero whole life carbon belief 2; Sustainable Outcomes Guide 9; UNSDGs (embracing) 9
ruination: embracing 21–22; retention 94–95; state 4
Ryú Peel (Ménard Dworkind Architecture & Design) 18, 20–21, 20f, 24
RYÙ Peel (Montreal, Canada) 1f

Sala Beckett International Drama Centre (*see also* Flores & Prats) 55, 65, 65f, 65–71; completion/location 66; entrance vestibule, enlargement 68; investigative drawing study analyses 69f; oval cuts 69–70; palimpsestic textual techniques 68f; project response, impact 70; ruinous narrative 66f; ruinous state 68; sustainability 70–71; strategy 70f
Scarpa, Carlo 43–44
scenography 112
Schmidt, Marian 127; biographic narrative possibilities 129
Schwitters, art 32f, 33
Scripto inferior (faint remnants of underwriting) 30
Scripto Superior (overwriting) 30
sculpting, new forms 105
Seah, Colin 60–62
Sedgwick Rd. (*see also* Olson Kundig Architects) 126, 144–149; ad-hoc angled props, usage 147; budget limitations 148; creative approach 146; found object possibilities 145; frames, impact 148; historic fragments, reuse/repurposing 144f; in situ reclamation 145f; interior, redrafting 148; narrative 145–146; palimpsestic reimagining 148; palimpsestic response 145; plan drawing 147f; project narrative 146; project synopsis 144–145; section drawing 147f; steel bracing, usage 147; sustainability 148–149; textual

techniques 146–148; unfinished business 148; wheeled partition walls ('Frank') 146–147, 147f
selective dismantling (Rotor) 109–111, 111f
Selo Shoe Shop (*see also* MNMA) 126, 127–131; cast furniture, usage 129; golden joinery, expression 130f; minimalistic approach 128–129; narrative 128–129; palimpsestic interior 127f; Prix Versailles Special Prize (award) 128; project narrative, delivery 128f; project synopsis 127–128; small-scale retail projects 127–128; sustainability 131; textual techniques 129–131; vernacular materials, usage 127–128
sensorial realm 160
sequential chronology, consideration 11
sequential interruptions 154
sequential motifs, materialisation 134
shared ownership 67
shared spaces, communal sensibility 91f
shared traits, highlighting 41
Shelley, Mary 35
Shophouses (Singapore), prevalence 62
site: chronological history, rewriting 151; efficacy, reinforcement 8; evolution, exploitation 118; incident (mapping), separation/clarification (usage) 130–131; listening 115–116; palimpsestic rereading 43; stories, accumulation 43–44
site-based stories, usage 52
site biography: extension 133f; notion 5
site narratives 5; enhancement/extension 54–55
site-sensitive approach, storytelling (impact/importance) 133–134
site specific narratives 4
solitude, experience 129
Sous rature (under erasure) 36
space: blank canvas treatment 20–21; creation 129; expressions 39–40; negative spaces, concrete cast 34;

shared spaces, communal sensibility 91f; transformation 97f; unexpected spaces, visual connections 94
spatial relationships, unwriting 89
specialisms 5
spolia, prevalence 7
stair treads, materiality 64
Star Machinery Building 145
Stone, Sally 43
storytelling 3; architectural storytelling 23, 42; categories 12; form, impact 17; impact/importance 133–134; palimpsest, relationship 1–2; primacy, instances 11; revelatory/revisionary storytelling 52; subversion 17; usage 11–12
structural stability, considerations 79
structural support, new forms 105
structure 97; consideration 11; undoing/decomposing/desedimenting 36
Studio Libeskind (*see also* the Military History Museum) 126, 150–155; architectural response 151
subtraction 88; palimpsestic techniques 114–115
subtractive approach 118
subtractive cuts, visual impact 100–101
subtractive incidents (fugues), control 99
subtractive methodology 160
subtractive preoccupation 88
subtractive process, integration 120
subtractive technique 162
succession, immersive dialogue 70
supply chains, collapse 159
surface articulation 53–54
surface decay, retention 94–95
surveying, new forms 105
sustainability 59, 64, 70–71; analysis 53; assessment tools 10; palimpsest, relationship 1–2, 7–10; role 52; strategy 70f
sustainable agenda, absence 17
Sustainable Outcomes Guide (RIBA) 9
sustainable thinking, embedding (ability) 5

Sustaining Specifying Guide, BIID carbon footprint report 2
symbiosis/interdependence, impact 40

Tabula rasa approach 44
Taha, Amin 12, 14–15
TAKK (*see also* the 10K House) 55, 161
temporal biographical approach 42–43
temporal boundaries, dissolution 38
temporal continuity/coexistence, achievement 64
temporal incidents, impact 88, 156
temporal interior interruptions, series 3
temporal interruptions, series 4
temporality, motif 7
temporal juxtaposition 24
temporal layering, process 41
temporal marriage 105
temporal relationship 30–31
temporal zone, palimpsest (presence) 30
textual analogies: connections 41; usage 52
textuality: examination 35; methodology 11–12
textual metaphors, usage 41
textual techniques 58–59, 62–64, 67–70; embracing 154; usage 47, 52, 53
theatrical company, site reoccupation 142
'The Day after House' 72
Theodoret of Cyrrhus 30
Thought by Hand 65
time: passage, importance 40; shadows, impact 64
Tompkins, Steve 138
translation: importance 36, 46; literal act 40
Turrell, James 129
Twiggy (project) 114–115

underlying text, uncovering 36
underwriting: 87–122 (chapter); evidential change 30–31; faint remnants 30; original underwriting 58–59; palimpsestic devices/techniques 53–54
Undoing Buildings (Stone) 43
undoing, care 100–101
United Nations 17 Sustainable Development Goals (UNSDGs) 9–10
unmaking (palimpsestic textual technique) 117f
unwriting: 87–122 (chapter); care 100–101; expression, temporal incidents (impact) 88; overwriting, combination 89; palimpsestic devices 53–54; palimpsestic interior 96f; palimpsestic language 99–100; palimpsestic technique 53–54, 105; process, usage 111f; rewriting/overwriting, conflation 124; subtractive process, integration 120; subtractive technique 162; technique 6, 102f, 120; usage 117f
upcycling 64
upgrading 6
urban decorative decay 21
urban narratives, changes 39
Urban Redevelopment Authority (URA), conservation district identification 62
usage value 71

vernacular materials, usage 127–128
Vertical Lane House project name, alternative 92
Vervoordt, Alex 46
vestige traces/fragments, notions 46
Villeglé, Jacques 33
violence, anthropological considerations 154
virgin materials, respect 59
visual slippages 92–93, 162–163

Wabi Sabi (Japanese philosophy): ideals 46; integration 21; kintsugi, resonance 130
Waldorf Hotel, The (pun) 17
Walled Off Hotel, The (Banksy) 12, 16–17, 16f

waste: recycling, EU targets 10; reduction, occurrence 9

Waterhouse at South Bund (*see also* Neri&Hu Design and Research Office) 89, 90–95; comfort/discomfort, voyeuristic interplay 93; erasure, palimpsestic textual techniques 93; host building, deletion 95; host building, response 94; indigenous archetype 92; mirrored window shutters 94; narrative 91–92; openings, approach 93f; partial unwriting, subtractive textual technique 90f; personal space, visual/aural/physical limitations 93; place/place-based heritage, examination 90–91; program/site/function/history, specificities 90; project synopsis 90–91; regional vernacular typologies 91, 92; retention, tempering 95; shared spaces, communal sensibility 91f; surface decay/ruination, retention 94–95; sustainability 94–95; textual techniques 92–94; triple-height entrance lobby 93–94; unexpected spaces, visual connections 94; visual slippages 92–93; visual slippages, approach 93f

wheeled partition walls ('Frank') 146–147, 147f

Whiteread, Rachel 15, 33–34

wholesale demolition 118–119

windowsills, recasting 117

Witherford Watson Mann Architects (WWM) 55, 162; projects, identification 79; proposal 79; ruin maintenance/habitation (*see also* Astley Castle) 81

World Trade Centre, redevelopment 150

World War II, aerial bombardment (results) 45–46

writing: palimpsestic devices/techniques 53–54; palimpsestic methodology 3

Zeitz, Jochen 104–105; adaptive reuse project 107

Zeitz MOCAA (*see also* Heatherwick Studio) 89, 102–113; location 103; narrative 109–110; project synopsis 102–105, 108–109; sustainability 112–113; textual techniques 105–107, 110–112

Zumthor, Peter 44